The Ultimate Chemical Equations Handbook

Student Edition

Second Edition

By

Jane D. Smith

and

George R. Hague, Jr.

FLINN SCIENTIFIC

"Your Safer Source for Science"

P.O. Box 219 • Batavia, IL 60510
(800) 452-1261 • Fax (866) 452-1436
Website: www.flinnsci.com • E-mail: flinn@flinnsci.com

ISBN 978–1–933709–30–7

Printed in the United States of America

Table of Contents

Exercises

About the Authors

Jane D. Smith

Jane Smith has more than 25 years of teaching experience with 20 of those years teaching Advanced Placement Chemistry. Jane graduated with a B.A. in Chemistry from Austin College in Sherman, Texas. After a short stint of research in the Biochemistry Department at the University of Texas Health Science Center in Dallas, she returned to Austin College to pursue certification in education. The result was an M.A. in Education and, in 1984, the beginning of a beloved teaching career at Van Alstyne High School in Van Alstyne, Texas, where Jane was the entire science department.

Jane taught chemistry at R. L. Turner High School in Carrollton, Texas for 19 years. In addition to the responsibilities of teaching chemistry and Advanced Placement Chemistry, she led a number of related activities over the years. Jane was an Odyssey of the Mind coach, a University Interscholastic League Academic Math and Science coach, Science Department Chair, and also served on a variety of school and district committees. Currently, she teaches chemistry and is the Math and Science coach for the Academic Decathlon team at Centennial High School in Frisco, Texas.

Jane is very active in the Science Teachers Association of Texas and the Associated Chemistry Teachers of Texas. She presents at many statewide workshops and has served as president of the Executive Board and newsletter editor for Associated Chemistry Teachers of Texas (ACT_2). In 1998 Jane received the Werner W. Schulz Award given by the Dallas–Ft. Worth Section of the American Chemical Society. In 1998–99 she was awarded an Honorable Mention in the Radio Shack Tandy Scholar Teachers program. In 1999 she received the Outstanding Chemistry Teacher Award from the Associated Chemistry Teachers of Texas. In 2003 she received the Southwest Regional Award in High School Chemistry Teaching from the American Chemical Society. When she is not teaching or working at school, Jane enjoys reading and traveling with her husband Chris.

George R. Hague, Jr.

George Hague, well known as a teacher, mentor, and valued colleague, passed away on July 15, 2002, after a long battle with leukemia. George was a 39-year teaching veteran with 22 years of experience as a New Jersey public school teacher and 17 years as a Texas private school teacher. He last taught College Prep Chemistry and AP Chemistry at St. Mark's School of Texas in Dallas, where he held the Leonard N. (Doc) Nelson Alumni Master Teaching Chair in Science. Born in East Orange, NJ, George received his B.A. in 1962 and M.A. in 1966 from Montclair State University, Upper Montclair, NJ. He did additional graduate work at Beaver, Cornell, Marquette, Rutgers, Seton Hall, NYU, Hope, and UC–Berkeley.

Hague was passionate about his love for chemistry, teaching, and his students. He has been quoted as saying, "If you are teaching chemistry and you are not having fun, then you must be doing something wrong!" George believed in "hands-on" science, demonstrations, music, and gimmicks that motivated in order to teach chemistry to his students. He was probably best known for his role as "Captain Chemistry," but had also assumed the identities of Merlin the Magician, "The Wizard of Chemistry," a Chemistry Cheerleader, and Albert Einstein. He presented hundreds of workshops and demonstrations at schools, colleges, convention centers, industrial locations, and even in parking lots. More than a hundred of his articles with tips for teachers appeared in such publications as *The Science Teacher, The Journal of Chemical Education, Chem 13 News, NJSTA Newsletter, The STATelite, ACT$_2$ Newsletter,* and *The Southwest Retort.* Awards won by George include the 1999 National Catalyst Award, Montclair State University Alumni Award, National Tandy Technology Scholar Award, AP Special Service Award from ETS, Outstanding Texas Chemistry Teacher of the Year Award, ACS Southwest and Middle Atlantic Regional Awards, Dreyfus Master Teacher, Shell Merit Teacher, Fellow of the New Jersey Science Teachers Association, Who's Who Among America's Teachers, and the 1983 Presidential Award for Excellence in Science Teaching.

George taught Advanced Placement Chemistry for more than 25 years—the first 12 at Bernards High School in New Jersey and the rest at St. Mark's School of Texas. George hosted chemistry camps and served five years as an AP Chemistry reader. His students competed successfully in state and regional scientific competitions and won four National TEAMS Championships in JETS competitions along with 10 Varsity State Championships and six Junior Varsity Championships.

George's professional activities included serving two terms on the NSTA Board of Directors (1985–88 and 1993–95), president of the New Jersey Science Teachers Association, president of the Associated Chemistry Teachers of Texas (ACT$_2$), member of the Board of Directors of the National Mole Day Foundation, and chairperson of the NSTA High School and Safety Committees.

George will be greatly missed.

Acknowledgments

This handbook has been a labor of love. This book is dedicated to all of the Advanced Placement Chemistry students we have ever taught. We thank them for inspiring us to become better teachers and continually restoring our faith in the youth of America. It has been both an honor and a privilege to be their chemistry teachers. *Note:* George wrote his acknowledgments in the fall of 2000.

I would like to thank the following individuals who have had a tremendous impact on molding me into the teacher and person I am today:

Elva Duran who ignited my love of science in the 5th grade and Rayburn Ray who kept the fires burning throughout my years at Coronado H.S., El Paso, TX.

Dr. Charles Barr and Dr. Michael Imhoff, chemistry professors at Austin College, Sherman, TX for inspiring me to pursue chemistry as a major and providing me with the opportunity to whet my appetite for teaching.

Lennie Sunthimer and Robyn Shipley-Gerko as well as so many other colleagues at R. L. Turner H.S., Carrollton, TX who encouraged and challenged me to be my best.

Carol Brown, Kristin Jones, Lisa McGaw, and Jerry Mullins, AP consultants who share generously with new AP teachers.

My family and especially my husband Chris, because he is my anchor and while he encourages me to say no, he supports me when I say yes.

Jane

I would like to thank the following individuals who have had a tremendous impact on molding me into the teacher and person I am today:

Irv Gawley, retired chemistry professor and early mentor, Montclair State University, Upper Montclair, NJ.

Warren Baecht, retired high school chemistry teacher and mentor, Tenafly H.S., Tenafly, NJ.

The late Hubert Alyea, master demonstrator and Princeton University professor.

George Gross, good friend and fellow chemistry teacher, retired from Union Township H.S., Union, NJ.

Patricia Lesinski Hague, my wife, best friend, soul mate, and mother of our three sons. A special thanks for putting up with all of my strange hours and professional activities.

Alice and George R. Hague, my parents!

George R. Hague Jr

Lastly, we wish to express our gratitude to Larry Flinn, III, Chairman of Flinn Scientific, who encouraged us to write this handbook.

"Chem is try!"

Introduction to Writing AP Equations

The Advanced Placement® (AP) Chemistry course is equivalent to the freshman General Chemistry course offered at the collegiate level. All AP Examinations are developed and administered by ETS on behalf of the College Board and are taken annually by high school students during the early part of May. The College Board sponsors the Advanced Placement® programs and decides how they will be constructed and used. Detailed information about the AP Program may be obtained at the College Board Web site: www.collegeboard.org/ap

The AP Chemistry Examination is a 185-minute examination, divided into two parts. The first part of the examination consists of 75 multiple-choice questions with a broad coverage of topics. Part One is 90 minutes in length and constitutes 50 percent of the final examination grade. The second part of the examination is the free response section. Part Two is 95 minutes in length and represents 50 percent of the final examination grade. For the first 55 minutes of the free-response section (Part Two), students are allowed to use a calculator and reference sheet as they work on three comprehensive problems. After 55 minutes have passed, calculators must be put away and may not be used for the remaining 40 minutes of the exam. During this final 40 minutes, students answer two essay questions as well as a descriptive chemistry question requiring the determination of products of chemical reactions (the Equations section).

Answering the free-response essay questions enables a student to demonstrate his or her ability to reason by applying chemical principles to solving problems. It is expected that answers will be clearly presented in a logical and coherent manner. One of the questions in Part Two—the Equations section—pertains to descriptive chemistry and asks the student to write ionic and molecular formulas for reactants and products in various chemical reactions.

In the Equations section of the exam, three sets of reactants are described and students must predict products, write a complete balanced equation, and answer a related question for each reaction. With the exception of the periodic table and reduction potentials chart, calculators and other aids are not allowed on the Equations section. The equations are of mixed types. The Equations section is worth 15 points and is equal to 15 percent of the free-response grade. It is for this section of the exam that this handbook was primarily written.

All AP equations "work." This means that in every case, a reaction will occur—there are no "no reaction" or equilibrium-type (double arrow/reversible) equations to worry about! All equations **must** be written as net ionic equations. Spectator ions are not included in net ionic equations and all other ions must be written in their ionic form. All molecular substances and insoluble compounds are to be written together, in their unionized form, as molecular formulas. Solids and pure liquids are also written using their molecular formulas. A saturated ionic solution is written in ionic form, while the ions in a suspension are written together (molecular form). It is also not necessary to include descriptive phase symbols such as (s), (l), (aq) and (g). This is a waste of valuable time as it will provide no extra points.

Each equation is worth a total of five points and is generally scored as follows: One point is given for correct reactants and two points for all correct products. If a reaction has three products, one point is given for two correct products and two points for all correct products. One point is awarded for balancing and one point for correctly answering the related question. Any spectator ions you write on the reactant side will nullify the one possible reactant point, but if they appear again on the product side, there is no product-point penalty. A fully molecular equation used for ionic substances will earn a maximum of one point. Charges shown on all ions must be correct!

The best method of preparing for the Equations section of the AP Exam is simply to practice a lot of equation-writing. The equation sets each year are similar and some equations show up year after year. When predicting a reaction, first try to classify it by type. For example, if the equation mentions acidic or basic solutions, it is most likely a redox reaction. Reactants that are combined with excess oxygen or burned indicate combustion reactions. During the practice process, if you find yourself totally at a loss, look up the reactants in this handbook, in the index of your textbook, in other reference books, or on the Internet to try to find information that will help you with the equation. Read this handbook and learn strategies for predicting various types of reactions. Do the practice problems in this book. Your AP Chemistry teacher will provide you with many AP type practice equations to work on. Many of the equations come from reactions that you will perform in the laboratory. Remember, all reactions do not fit neatly into the five types of equations generally learned in first year chemistry courses. Save the equation sheets that you work on and practice them again, just before the AP exam in May.

The types of equations most often encountered on the AP Chemistry Examination usually fit into one (or more) of the following categories:

- Acid–Base
- Addition
- Anhydrides
- Combustion
- Complex Ion
- Decomposition

- Double Replacement (Metathesis)
- Electrolysis
- Organic
- Oxidation–Reduction (Redox)
- Precipitation
- Single Replacement

Solubility rules as well as the strong acids and strong bases **must** be memorized if one is to be successful at predicting reactions. Rules for writing formulas and the charges and names of common polyatomic ions must also be memorized. How to get formulas and charge information from a periodic table is a most helpful skill to learn. A periodic table and a table of standard reduction potentials are provided in the AP Chemistry Examination booklet for use during the free-response section of the examination.

The Number of Equations Used on the AP Chemistry Examination is Finite!

Practicing Equations Will Help to Improve Your AP Score!

Chapter 1
A Brief Look at the Symbols and Nomenclature of the Elements

"Elementary Speaking, Mr. Spock, It Is a Logical System!"

Mr. Spock, the Science Officer and second-in-command of the Starship Enterprise, gained great fame in the movies and television series known as *Star Trek*. Spock was a Vulcan. Vulcans, according to the *Star Trek* series, were known for not having emotions and for having extremely logical minds.

Before the untimely demise of *Star Trek*, it was rumored that Mr. Spock questioned the supposedly illogical earthling system of naming the elements, the building blocks of matter. After all, four elements had been named after the town of Ytterby, Sweden, (that's right, four elements!); and others had been named after the planet Uranus, a budding twig, the state of California, a father and his daughter, a subterranean gnome, and Albert Einstein. What logic could possibly exist in such a system? The system appears to be based on a mixture of unrelated facts that have absolutely nothing at all to do with chemistry!

But wait, Mr. Spock, there is earthling logic to the naming of the elements. The discovery or creation of a new element must be verified by other scientists and then must be accepted by the International Union of Pure and Applied Chemistry (IUPAC). The new element must then be named. Certain rules apply, for example, metals names are supposed to use the suffix "ium." Yet for the most part, the discoverer chooses the name. In other words, if you were to discover a new element, you would have the right to name it— you may even name it after your high school, or your hometown!

The names of the known 117 elements (only the first 112 have official names) fit into one or more of the following eight categories:

- Properties of the elements
- Sources of the elements
- Countries and regions of the world
- Cities
- Heavenly bodies
- Mythological individuals
- Individuals
- Miscellaneous names—in most cases, names applied to some of the elements discovered by the ancients.

For Mr. Spock's benefit, let us take a look at uranium and a number of the transuranium elements that were discovered at the University of California–Berkeley. Uranium was discovered just after the discovery of the planet Uranus. The element was named in honor of the new planet.

In 1940, when element 93 was first prepared at the University of California–Berkeley's Lawrence Radiation Center, it only seemed logical to name the new element next to uranium after the planet next to Uranus, which at that time was Neptune. The same line of reasoning was used to name element number 94. Because the planet beyond Neptune was Pluto, the name plutonium resulted. How is that for logic, Mr. Spock? (Of course, now Pluto has lost its claim to planethood, but plutonium persists.)

Since many of the transuranium elements were prepared at the Lawrence Radiation Center, University of California at Berkeley, California, U.S.A., it is logical to find the following names:

Americium—named after the Americas (after all, one must be patriotic!). Furthermore, americium is in the same family (homologs) as europium, which is also a continent (logic strikes again).

Californium—named after the state and university where it was first discovered.

Berkelium—named after the city where it was first synthesized.

Lawrencium—named after the Lawrence Radiation Center. Ernest O. Lawrence was a professor at the University of California–Berkeley where he invented the cyclotron.

Now let's look at some names of elements that have been logically derived from their properties:

Actinium—from the Greek word *aktis* which means beam or ray. Actinium is an alpha ray emitter with a very short half-life.

Argon—from the Greek word *argon* meaning inactive.

Bromine—from the Greek word *bromos* for stench, a perfect description of its suffocating odor.

Hydrogen—comes from two Greek words *hydro* and *genes* that together mean "water former."

Astatine—from the Greek word *astatos* or unstable. Astatine is a radioactive element with a very short half-life.

Mercury—from the Greek word *hydrargyrum* for "quick silver" or "liquid silver."

Radium—from the Latin word *radius* for ray. It is highly radioactive and glows in the dark.

Many other elements have names relating to their colors. Five are named from the colors they emit when they are burned and their spectra are observed through a spectroscope:

Helium—from the Greek word *helios* or "sun," where its first spectral lines were discovered.

Cesium—from the Latin word *caesius* for "sky blue."

Rubidium—from the Latin word *rubidius* for the dark red colors emitted in its excited state.

Indium—named for its indigo-blue spectrum.

Thallium—from the Latin word *thallos* meaning "budding twig," "young shoot" or "green branch," a very logical description of the green line observed in its spectrum. *Note:* This is your authors' favorite name for an element.

Other elements have been named from the properties of their salts:

Aluminum—from the Latin word *alumen* or alum, which has an astringent taste.

Boron—derived from the Arabic word *buraq*, which refers to the white color of borax.

Chromium—from the Greek word *chroma* or *chromos* meaning color; many of chromium's salts are colored.

Iodine—from the Greek word *iodes* for violet, the color of its vapor when iodine sublimes.

Lanthanum—from the Greek term *lanthanein*, which means concealed. It is very difficult to separate lanthanum from other rare earth elements.

Praseodymium and **neodymium** were discovered together. Praseodymium (Greek: *prasios* and *didymos*) means "green twin" while neodymium (Greek: *neos* and *didymos*) means "new twin."

Osmium—from the Greek word *osme* or odor; the element is a metal with a pungent odor.

Rhodium—from the Greek word *rhodon* meaning rose; its salts form rose-colored solutions.

We hate to admit it, Mr. Spock, but some elements received their names from "mistakes" or "errors." We are sure you will find the following names to be unforgivable:

Manganese—comes from the Latin word *magnes* or "magnet." Unfortunately, its ore was first confused with magnetic iron ore.

Nickel—from the Swedish abbreviation for *kopparnickel* or "false copper." This was the result of confusing a reddish ore that looked like copper ore but was found to contain nickel instead.

Oxygen—this is the all-time "mistake" in terms of naming an element. Antoine Lavoisier, "The Father of Modern Chemistry," made the mistake. Oxygen comes from two Greek words (*oxys* and *genes*) and means "acid former." With logic like that, hydrochloric acid could not possibly exist! Score one for Mr. Spock!

Elements named after their sources include the following:

Calcium—from the Latin word *calx* or "lime," which is now known as calcium oxide.

Carbon—from the Latin word *carbo*, which means coal or charcoal.

Potassium—from the Latin word *kalium*, which means potash (potassium carbonate).

Sodium—from the Latin word *natrium* or soda (sodium carbonate).

Tellurium—from the Latin word *tellus*, which means earth and describes where it is found in nature.

A number of countries and regions in the world have given their names to the elements. Many of these elements were named by patriotic individuals in honor of their homelands or region of origin:

Francium—France.

Germanium—Germany.

Polonium—Poland.

Gallium—from the old Latin name *Gallia* for France.

Rhenium—from the old Latin name for the Rhine provinces in Germany.

Scandium—Scandinavia.

Thulium—from *Thule*, the ancient name for Scandinavia.

Ruthenium—from the Latin name *Ruthenia* for Russia.

Hassium—from *Hassia*, the Latin name for the German state of Hesse, the home of a large German nuclear research facility. The actual location is Darmstadt, Germany in the state of Hesse.

Under cities we find the "King of them all," Ytterby, Sweden! Four elements were discovered in the rich mineral fields near the small town of Ytterby, and were named in its honor. These elements include **erbium**, **terbium**, **ytterbium**, and **yttrium**. Once again, logic prevails!

Other cities with elements named after them include:

Copenhagen, Denmark—has the Latin name *Hafnia*, hence the element **hafnium**. Interestingly enough, neither of its codiscoverers was Danish. The element was named by Dick Coster, who was Dutch, and George Charles de Hevesy, who was Hungarian, while both were connected with the Neils Bohr Institute of Theoretical Physics in Copenhagen.

Holmia, Sweden—the home city of the discoverer of **holmium**.

Paris, France—this city has given its name to **lutetium** since *Lutetia* is the ancient name of the city. A logical guess will give you the hometown of Georges Urbain, the discoverer of lutetium.

Strontian, Scotland—the site where the element **strontium** was first found. Once again, Mr. Spock, we have more logic for you to ponder!

Dubna, Russia—the element **dubnium** was named in honor of the Joint Institute for Nuclear Research in Dubna, Russia.

Darmstadt, Germany—the element **darmstadtium** was named after the place of its discovery.

We have already discussed a number of elements named after astronomical objects. Two others are worthy of mention. Both were discovered at approximately the same time as astronomical discoveries and, as a result, were named in their honor.

Cerium—after the asteroid Ceres.

Palladium—from the asteroid Pallas.

A number of mythological individuals have given their names to elements.

Vanadium, niobium, and **tantalum**—all are found in the same family (homologs). Niobium is always found in nature with tantalum. Tantalum derives its name from the mythological Greek king known as Tantalus. His daughter, Niobe, has given her name to niobium. Just like the elements, the father and daughter are closely related. The other member of the family, vanadium, was named after the Scandinavian goddess Vanadis.

Promethium—the first artificial element produced in a laboratory. It was named after Promethius who stole fire from the heavens for the use of mankind. The analogy is that the element promethium was prepared (stolen) by harnessing nuclear fission.

Titanium—one of our modern-day super metals; it was named after the "supermen" in Greek mythology known as the Titans.

Cobalt—a very interesting name derived from the German word *kobald* for a subterranean gnome, goblin or evil spirit. The name came into being because the poisonous ores of cobalt were very treacherous for miners to mine.

Several elements have been named in honor of famous scientists.

Curium—named in honor of Pierre and Marie Curie for their investigations in radioactivity.

Einsteinium—named after the brilliant physicist Albert Einstein, *Time* magazine's "Man of the Twentieth Century."

Fermium—honors the great Italian nuclear physicist Enrico Fermi.

Gadolinium—first isolated from the mineral *gadolinite*. Gadolinite was named after John Gadolin, a chemist from Finland.

Mendelevium—named after Demitri Mendeleev, the Russian scientist who is considered to be the "Father of the Periodic Table of the Elements."

Nobelium—honors Alfred Nobel, a Swedish scientist, who was the inventor of dynamite. He was also responsible for establishing the Nobel Prizes.

Samarium—this element gets its name from the mineral *samarkite* from which the element was first discovered. Samarkite was named after a Russian engineer and mine official by the name of Colonel V. E. Samarsky.

Rutherfordium—named after Lord Ernest Rutherford who was born in New Zealand and did his research in England. There he developed the nuclear theory of the atom and discovered the proton; he is considered to be the founder of nuclear physics.

Seaborgium—named after the great American nuclear chemist Glenn T. Seaborg, who was responsible for synthesizing many of the transuranium elements (those beyond uranium) now found on the periodic table. Seaborium is the only element ever named after a person who was actually alive at the time of the official naming. Such a practice had been in conflict with IUPAC policies prior to that time.

Bohrium—named after Niels Bohr, the Danish physicist who formulated the quantum theory of the electronic structure of the hydrogen atom and the origin of the spectral lines of hydrogen and helium.

Meitnerium—named after the great Austrian female physicist, Lise Meitner. Her discoveries in nuclear physics played a major role in developing nuclear energy.

Roentgenium—named after German physicist, Wilhelm Conrad Roentgen. His study of cathode rays led to his discovery of x-rays in 1895 and the Nobel Prize in Physics in 1901.

Copernicium—honors Nicolaus Copernicus, a scientist and astronomer who lived from 1473 to 1543. His heliocentric model of the universe showed that the Earth orbits the Sun, changing the way the world was viewed.

Mr. Spock will surely point out that some of the ancient elements such as antimony, iron, lead, sulfur, tin, and zinc are not very logically named. The names of those elements have been derived from old words and their original meanings have become obscured with time. But fear not, Mr. Spock, a *logical* system for the preliminary naming of elements without official names has been developed by the IUPAC Commission on the Nomenclature of Inorganic Chemistry.

The IUPAC Commission has made the following recommendations for the unofficial names of new elements:

- Names should be short and obviously related to the atomic numbers of the elements.
- Names should end in "ium" for both metals and nonmetals.
- Symbols for these systematically named elements should consist of three letters.
- Symbols should be derived from the atomic number and be related visually, as much as possible, to their names.

The three-letter system was developed in order to avoid duplication of already existing two-letter symbols. The following roots are used to derive the names of new elements:

0 = nil
1 = un (pronounced with a long u to rhyme with "moon")
2 = bi
3 = tri
4 = quad
5 = pent
6 = hex
7 = sept
8 = oct
9 = enn

To name an element, the roots are put together in the order of the digits found in the atomic number of the element, and the suffix "ium" is added to the name. According to the system, element #118 would be called ununoctium and its symbol would be Uuo since the first letter of each root in the name is used. The final "i" in bi or tri is dropped when these letters appear before "ium." The final "n" in enn is dropped when it appears before nil.

Atomic Number	IUPAC Name	Symbol
113	ununtrium	Uut
114	ununquadium	Uuq
115	ununpentium	Uup
116	ununhexium	Uuh
117	ununseptium	Uus
118	ununoctium	Uuo

Mr. Spock, is there a glimmer of a smile on that unemotional face of yours? Let us warn you, however, that the discoverers of a new element will continue to have the right to suggest another name to the IUPAC Commission once evidence or proof of their discovery has been accepted by the scientific community.

Let us, just for a moment, imagine what it would be like if the IUPAC alternative "root" system were expanded to cover all of the elements. Water would be composed of the elements unium and octium. Table salt would contain ununium and unseptium. FORGET IT! Give us the old system of naming elements! Maybe Mr. Spock is right, our method of naming elements is not perfectly logical. As things now stand, the names of the first 112 elements tell 112 different stories. These stories tell about "The Romance of Chemistry" and the humanizing of science. This is something logical Mr. Spock might never understand, but as for us, we are ever thankful for the humanizing of chemistry!

Determine the formula for EbE_6 using the IUPAC alternative "root" system. Follow the two examples given below.

Example 1: U_2O = unium octium

                                 = atomic   atomic
                                     #1   #8

                                 = H_2O

Example 2:   UuUs   = ununium unseptium

                                 = atomic   atomic
                                     #11   #17

                                 = NaCl

Practice Problem: EbE_6 = ????

Exercise 1–1: Symbols and Formulas

Chemical symbols are used for convenience to represent the names of the elements. Some symbols are merely initials of the names. All of these one-letter element symbols are for older elements. No new elements that are discovered in the future will be allowed to have symbols with only one letter.

H _____	O _____	C _____
F _____	I _____	B _____
U _____	P _____	N _____
Y _____	S _____	V _____

Other element symbols consist of the first letter in the name of the element and a second letter which is prominent in pronouncing the name of the element. (This is the modern day rule that is followed.)

Si _____	Ca _____	Bi _____
Ra _____	Al _____	Mn _____
Cl _____	Cd _____	Br _____
Li _____	Ni _____	Zn _____
Mg _____	As _____	Ba _____
Pt _____	Pu _____	Np _____
Co _____	Ti _____	Cr _____
Sg _____	Sr _____	Ga _____

Some symbols are derived from non-English words, that is, Latin, Greek or German names.

Fe	(ferrum) _____	Cu	(cuprum) _____
Na	(natrium) _____	K	(kalium) _____
Ag	(argentum) _____	Hg	(hydrargyrum) _____
Sn	(stannum) _____	Sb	(stibium) _____
Pb	(plumbum) _____	Au	(aurium) _____
W	(wolfram) _____		

Symbols are used as a sort of shorthand in writing the names of elements. The use of symbols to represent atoms, or definite quantities by mass of the elements, is also important in writing chemical formulas and in describing reactions. Thus, the symbol C represents the element carbon, but it also represents one atom of carbon. *Note:* The first letter of the symbol is **always** printed uppercase; the second letter is **always** printed lowercase!

Fill in the blanks above with the names of the elements the symbols represent. Be able to give either the symbol or the name of the element from memory.

Chapter 2
Simple Inorganic Formulas and Nomenclature

Compounds consisting of two different elements in various ratios are considered to be *binary compounds*. Binary compounds usually end in the suffix "ide." There are two types of binary compounds—binary molecules and binary salts. A binary molecule consists of two nonmetals bonded via covalent bonding. A binary salt consists of a metal and a nonmetal exhibiting ionic bonding.

General Rules

A. Binary Molecules (Nonmetal + Nonmetal) i.e., CO_2 or N_2O_3

Molecules are formed when two nonmetals or metalloids combine and prefixes must be used to designate the number of atoms of each element present in one molecule. Nonmetals are found just to the right of the zigzag line on the periodic table. Metalloids are near the zigzag line and may have some properties of metals and other properties of nonmetals.

Prefixes are used to designate the number of atoms of each element present in the formula of a binary compound. The prefix *mono* is **never** used in front of the first element (standard convention). If there is only one atom, the mono is assumed.

1 = mono	4 = tetra	7 = hepta	10 = deca
2 = di	5 = penta	8 = octa	11 = undeca
3 = tri	6 = hexa	9 = nona	12 = dodeca

Name the following binary molecules — CO_2 and N_2O_3

To determine the first word in the name of the compound:

1. Give the prefix designating the number of atoms of the first element present. Remember, *mono* is never used (by standard convention) for the first element.

   CO_2: No prefix for C N_2O_3: **di**

2. Name the first element.

   CO_2: **carbon** N_2O_3: **dinitrogen**

9

To determine the second word in the compound's name:

  3. Give the prefix designating the number of atoms of the second element present.

      CO_2: carbon **di** N_2O_3: dinitrogen **tri**

  4. Name the root of the second element. *Note:* The root is the base name that designates the element.

      CO_2: carbon di**ox** N_2O_3: dinitrogen tri**ox**

  5. Add the suffix *–ide* to the root of the second element.

      CO_2: carbon diox**ide** (official name) N_2O_3: dinitrogen triox**ide** (official name)

B. Binary Salts (Metal + Nonmetal) i.e., CaCl$_2$

Prefixes giving the number of atoms of each element present are *never* used to name an ionic salt. Salts exhibit ionic bonding between a metal and a nonmetal, while molecular substances exhibit covalent bonding between two nonmetals.

Name the following binary salt — CaCl$_2$

By convention, the metal is written before the nonmetal. To identify the first word in the name:

  1. Name the first element (metal).

      $CaCl_2$: **calcium**

To determine the second word in the name of the compound:

  2. Name the root of the second element (nonmetal).

      $CaCl_2$: calcium **chlor**

  3. Add the suffix *–ide* to the root of the second element.

      $CaCl_2$: calcium chlor**ide**

Exercise 2–1: In column 1, classify each of the following compounds as binary molecules (M) or binary ionic salts (I). Then in column 2, use the rules to name each binary compound.

1. CaF_2	_____ _____	10. SrI_2	_____ _____
2. P_4O_{10}	_____ _____	11. CO	_____ _____
3. K_2S	_____ _____	12. Cs_2Po	_____ _____
4. NaH	_____ _____	13. $ZnAt_2$	_____ _____
5. Al_2Se_3	_____ _____	14. P_4S_3	_____ _____
6. N_2O	_____ _____	15. $AgCl$	_____ _____
7. O_2F	_____ _____	16. Na_3N	_____ _____
8. SBr_6	_____ _____	17. Mg_3P_2	_____ _____
9. Li_2Te	_____ _____	18. XeF_6	_____ _____

Chapter 3
Oxidation Numbers: Anions and Cations

Metals with Variable Charges (Oxidation Numbers)

A number of metallic elements can form compounds in which the metal ions (cations) may have different charges. These charges are known as oxidation numbers and are sometimes referred to as valences. The transition metals in the middle of the periodic table have variable oxidation numbers as do many of the representative elements in groups 13–16 in the periodic table. Cations with variable oxidation numbers use a Roman numeral enclosed in parentheses to designate the charge on the metal ion. This naming system is called the Stock System. For example, the oxidation number of iron in the following two compounds cannot be the same: $FeCl_2$ and $FeCl_3$. Calling both of these compounds iron chloride would only lead to confusion. The Stock System is used to differentiate between ions that have two or more possible charges. $FeCl_2$ is known as iron(II) chloride and $FeCl_3$ is officially called iron(III) chloride. The Roman numeral represents the charge on the metal cation and does *not* represent the number of atoms of the element present. To name these types of ionic compounds, the oxidation numbers of all the elements present must be known.

Here are some simple rules that should help in the determination of the oxidation numbers of metallic ions (cations) from the formulas of their compounds.

1. The oxidation number of any **element** in its free state (uncombined with other elements) is zero, e.g., Fe in a bar of iron is zero. O_2 and N_2 in the Earth's atmosphere both have oxidation numbers of zero. When an element has equal numbers of protons and electrons, its overall charge is zero.

2. The oxidation number of **alkali metals** in a compound is always 1+, e.g., Li^+, Na^+, K^+, etc.

3. The oxidation number of **alkaline earth metals** in a compound is always 2+, e.g., Mg^{2+}, Ca^{2+}, Sr^{2+}, etc.

4. **Fluorine** is always assigned an oxidation number of 1^- in a compound, e.g., F^-.

5. The oxidation number of **oxygen** is almost always 2^- in a compound. Exceptions to this rule would be peroxides, O_2^{2-} where the oxidation number of each oxygen is 1^-, and superoxides, O_2^- where the oxidation number of each oxygen is $\frac{1}{2}^-$. Neither peroxides nor superoxides are common. Peroxides are only known to form compounds with the elements in the first two columns of the periodic table, e.g., H_2O_2, Na_2O_2, CaO_2, etc. Potassium, rubidium, and cesium are the only elements that form superoxides, e.g., KO_2. *Note:* The name superoxide may also be called superperoxide.

6. In covalent compounds with nonmetals, **hydrogen** is assigned an oxidation number of 1+, e.g., HCl, H_2O, NH_3, CH_4. The exception to this rule is when hydrogen combines with a metal to form a **hydride**. Under these conditions, which are rare, hydrogen is assigned an oxidation number of 1^-, e.g., NaH.

7. In **metallic halides** the halogen (F, Cl, Br, I, At) always has an oxidation number equal to 1^-.

8. Sulfide, selenide, telluride, and polonide are always 2^- in binary salts.

9. Nitrides, phosphides, and arsenides are always 3^- in binary salts.

10. All other oxidation numbers are assigned so that the sum of the oxidation numbers of each element equals the net charge on the molecule or polyatomic ion. In a neutral compound, the sum of the positive and negative charges must always equal zero.

11

Example

Determine the oxidation number of the underlined element: K$\underline{Mn}$O$_4$. Since K is an alkali metal, its charge must be 1+. Oxygen is 2– but there are four of them, therefore, 4 times 2– equals 8–. If 1+ and 8– are added together, we get 7–. In order for the compound to be neutral, the Mn must be 7+.

Algebraically, $(1+) + (x) + 4(2-) = 0 \quad \therefore \quad x = 7+$

Other Examples

$\underline{N}H_4^+$: The sum of the charges on this polyatomic ion must equal 1+. Since hydrogen has a 1+ charge and there are four hydrogen atoms, the nitrogen must be 3– because (3–) + (4+) = 1+!

$K_2\underline{Cr}_2O_7$: Potassium is 2 times 1+ = 2+, and oxygen is 7 times 2– = 14–. (14–) + (2+) = 12–. Since there are two chromium atoms and the compound is neutral overall, the charge on the two chromium atoms must be equal to 12+ and each chromium atom must have a charge of 6+ (since 12+/2 = 6+).

Algebraically, $2(1+) + (2x) + 7(2-) = 0 \quad \therefore \quad x = 6+$

$\underline{O}_2$: This is an element in its free state, so the oxidation number must be zero.

Note: Ions written alone, such as peroxide, must be written with a charge on them, e.g., O_2^{2-}. In a compound, the charges on individual atoms or ions are not shown.

Exercise 3–1: Determine the oxidation number of each underlined element.

1. $K_2\underline{S}$

2. $Na\underline{Cl}O_4$

3. $\underline{Br}Cl$

4. $Li_2\underline{C}O_3$

5. $\underline{O}F_2$

6. $\underline{S}_8$

7. $\underline{Mg}$

8. $K_2\underline{W}_4O_{13}$

9. $Mg(\underline{B}F_4)_2$

10. $\underline{Au}_2O_3$

11. $\underline{C}_{60}$

12. $\underline{Zr}O_2$

13. $\underline{Nb}OF_6^{3-}$

14. $Al_2(\underline{Cr}O_4)_3$

15. $Cs_2\underline{Te}F_8$

Remember, free elements, no matter how complex the molecule, have an oxidation number (valence or charge) equal to zero. The following are diatomic or polyatomic elements in nature which must be committed to memory. These elements exist as neutral molecules in nature!

Polyatomic Elements

Hydrogen, H_2	Bromine, Br_2
Nitrogen, N_2	Iodine, I_2
Oxygen, O_2	Ozone, O_3
Fluorine, F_2	Phosphorus, P_4
Chlorine, Cl_2	Sulfur, S_8

Most common forms of buckminsterfullerenes (buckyballs): C_{60} and C_{70}

Representative Elements (s- or p-block) Cations and Anions

Charges can be determined by position (family) on the Periodic Table. Cations (+ ions) come from metals that lose electrons (oxidation) in order to become isoelectronic with a noble gas. Anions (– ions) come from nonmetals that gain electrons (reduction) to become isoelectronic with a noble gas.

Oxidation Numbers (Valence) of Representative Element Cations and Anions							
1+ Alkali metals	2+ Alkaline earth metals	3+		4–	3– Nitrogen family	2– Oxygen family	1– Halogens
Lithium Sodium Potassium Rubidium Cesium Francium Hydrogen	Magnesium Calcium Strontium Barium Radium Beryllium	Aluminum Boron		Carbide	Nitride Phosphide Arsenide	Oxide Sulfide Selenide Telluride Polonide	Fluoride Chloride Bromide Iodide Astatide

More on Metallic Elements with Variable Oxidation Numbers

Transition metals, representative metals with p and d sublevels, and the inner transition metals typically have more than one oxidation state in compounds. Electrons for these metallic elements are lost (oxidized) from their outermost energy levels in the following order: p, s, d. Such elements are *not* isoelectronic with a noble gas when the outermost (valence) electrons are lost and if enough energy is available, will begin to lose d level electrons.

Example 1: A neutral vanadium atom has an electron configuration of [Ar] $4s^2\ 3d^3$. The outermost electrons are always lost first, therefore, vanadium will lose its $4s^2$ electrons and form the vanadium(II) ion, V^{2+}. With additional energy, the V^{2+} cation can lose its $3d^3$ electrons in order, forming vanadium(III), V^{3+}, vanadium(IV), V^{4+}, and vanadium(V), V^{5+} cations.

Example 2: The electron configuration for an atom of Fe is [Ar] $4s^2\ 3d^6$. The first cation that forms when the $4s^2$ electrons are lost is the iron(II) ion, Fe^{2+}. Additional energy will cause the iron(II) ion to lose one of its 3d electrons to form the iron(III) ion, Fe^{3+}. The remaining d electrons are all spinning in the same direction and the energy required to oxidize them is greater than normally encountered in an ordinary chemical reaction. The repulsive forces between the only two paired electrons in the 3d sublevel make the formation of the iron(III) ion relatively easy.

Example 3: The electronic configuration of a neutral lead atom is [Xe] $6s^2\ 4f^{14}\ 5d^{10}\ 6p^2$. The two common oxidation numbers of lead are lead(II) when the two $6p^2$ electrons are lost and lead(IV) when the two $6s^2$ electrons are also oxidized. Tin behaves in a similar manner when it forms tin(II) and tin(IV) cations. Bismuth with an electron configuration of [Xe] $6s^2\ 4f^{14}\ 5d^{10}\ 6p^3$, forms bismuth(III) and bismuth(V) ions.

Inner transition elements are sometimes called by such names as the lanthanides, actinides, rare earth elements, and the transuranium elements. All of these elements are quite rare, and many of the elements beyond uranium (the transuranium elements) exist for only short periods of time. Reactions involving such elements are seldom encountered in a beginning chemistry course and there is little need to pursue this topic in any detail. Two inner transition elements worth mentioning are uranium (U^{3+}, U^{4+}, and U^{5+}) and cerium (Ce^{3+} and Ce^{4+}).

Both inner transition and transition elements are known for their variable oxidation numbers. The most common oxidation number for transition elements is 2+. The d sublevel in transition elements is responsible for the various oxidation numbers that result. Incomplete d sublevels are also responsible for the many colorful transition compounds that are known to exist. Complete d sublevels in cations of silver and zinc result in white compounds.

Summary of Cations with Variable Oxidation Numbers—Stock System

1+, 2+	copper(I), Cu^+; copper(II), Cu^{2+}; mercury(I)*, Hg_2^{2+}; mercury(II), Hg^{2+}

Note: mercury(I) actually exists as a diatomic ion and is written as Hg_2^{2+} and not Hg^+.

1+, 3+	gold(I), Au^+; gold(III), Au^{3+}; indium(I), In^+; indium(III), In^{3+}; thallium(I), Tl^+; thallium(III), Tl^{3+}
2+, 3+	chromium(II), Cr^{2+}; chromium(III), Cr^{3+}; cobalt(II), Co^{2+}; cobalt(III), Co^{3+}; iron(II), Fe^{2+}; iron(III), Fe^{3+}; manganese(II), Mn^{2+}; manganese(III), Mn^{3+}
2+, 4+	lead(II), Pb^{2+}; lead(IV), Pb^{4+}; platinum(II), Pt^{2+}; platinum(IV), Pt^{4+}; tin(II), Sn^{2+}; tin(IV), Sn^{4+}; zirconium(II), Zr^{2+}; zirconium(IV), Zr^{4+}
3+, 4+	cerium(III), Ce^{3+}; cerium(IV), Ce^{4+}
3+, 5+	antimony(III), Sb^{3+}; antimony(V), Sb^{5+}; arsenic(III), As^{3+}; arsenic(V), As^{5+}; bismuth(III), Bi^{3+}; bismuth(V), Bi^{5+}; phosphorus(III), P^{3+}; phosphorus(V), P^{5+}
2+, 3+, 4+	iridium(II), Ir^{2+}; iridium(III), Ir^{3+}; iridium(IV), Ir^{4+}; titanium(II), Ti^{2+}; titanium(III), Ti^{3+}; titanium(IV), Ti^{4+}
2+, 4+, 5+	tungsten(II), W^{2+}; tungsten(IV), W^{4+}; tungsten(V), W^{5+}
3+, 4+, 5+	uranium(III), U^{3+}; uranium(IV), U^{4+}; uranium(V), U^{5+}
2+, 3+, 4+, 5+	vanadium(II), V^{2+}; vanadium(III), V^{3+}; vanadium(IV), V^{4+}; vanadium(V), V^{5+}

Note: When reading the name of an ion such as Pb^{2+}, the ion is read in English as the "lead two ion."

Special Metallic Cations

The following transition metal cations do not exhibit variable oxidation numbers and are normally written without Roman numerals:

cadmium, Cd^{2+} silver, Ag^+ zinc, Zn^{2+}

Nickel, on the other hand, has variable oxidation numbers, and even though it almost always appears as the nickel(II) ion, Ni^{2+}, the Roman numeral must be written.

The ions of the representative elements gallium, germanium, and indium do not have variable oxidation numbers, but are written with Roman numerals:

gallium(III), Ga^{3+} germanium(IV), Ge^{4+} indium(III), In^{3+}

Polyatomic Ions

The term polyatomic ion is used to describe a group of atoms that behave as a single ion. The bonding within a polyatomic ion is covalent, but because there is always an excess or shortage of electrons when compared to the number of protons present, an ion results. A common polyatomic positive ion (cation) is the ammonium ion, NH_4^+. A common polyatomic negative ion (anion) is the sulfate ion, SO_4^{2-}.

Remember that polyatomic ions stay together as a group. The ammonium ion is always written as NH_4^+ and *never* as $N^{3-} + 4H^+$ or H_4^+ or H_4^{4+}. If two or more of the same polyatomic ions are needed within a compound in order to reach electrical neutrality, the polyatomic group is enclosed in parentheses. For example, ammonium sulfate is written as $(NH_4)_2SO_4$. The compound consists of two ammonium ions and one sulfate ion. The letters are read as "N, H, four taken twice, S, O, four."

Polyatomic ions must be memorized! There is no simple way to learn all of these ions but it is helpful to realize that some of them come in related pairs. For example, sulfate, SO_4^{2-}, and sulfite, SO_3^{2-} share the same charge and include the same elements, S and O, but they differ in their number of oxygen atoms. Notice that the *–ate* form has one more oxygen atom that the *–ite* form; in other words sul<u>fate</u> "ate" one more O than sul<u>fite</u>. There are several of these pairs, so if you know nitrate is NO_3^-, then it's easy to deduce that nitrite is NO_2^-. Chlorate is ClO_3^- and chlorite is ClO_2^-.

Another helpful tip is to observe patterns in the *–ate* formulas and their relationship to the periodic table. Notice that all of the *–ate* ions on the outside of the bold line have three oxygen atoms and the *–ate* ions on the inside of the bold line have four oxygen atoms.

	CO_3^{2-}	NO_3^-		
MnO_4^-		PO_4^{3-}	SO_4^{2-}	ClO_3^-
CrO_4^{2-}			SeO_4^{2-}	BrO_3^-
				IO_3^-

Common Polyatomic Ions

Anions

1–

acetate, CH_3COO^-

amide, NH_2^-

azide, N_3^-

benzoate, $C_6H_5COO^-$

bromate, BrO_3^-

chlorate, ClO_3^-

chlorite, ClO_2^-

cyanate, OCN^-

cyanide, CN^-

dihydrogen phosphate, $H_2PO_4^-$

formate, $HCOO^-$

hydrogen carbonate, HCO_3^-
   (bicarbonate)

hydrogen sulfate, HSO_4^-
   (bisulfate)

hydrogen sulfide, HS^-
   (bisulfide or hydrosulfide)

hydroxide, OH^-
   (called hydroxyl when aqueous)

hypochlorite, ClO^-

iodate, IO_3^-

nitrate, NO_3^-

nitrite, NO_2^-

perchlorate, ClO_4^-

permanganate, MnO_4^-

thiocyanate, SCN^-
   (thiocyanato)

triiodide, I_3^-

vanadate, VO_3^-

2–

carbide, C_2^{2-}
   (saltlike)

carbonate, CO_3^{2-}

chromate, CrO_4^{2-}

dichromate, $Cr_2O_7^{2-}$

imide, NH^{2-}

manganate, MnO_4^{2-}

metasilicate, SiO_3^{2-}

monohydrogen phosphate, HPO_4^{2-}

oxalate, $C_2O_4^{2-}$

peroxide, O_2^{2-}

peroxydisulfate, $S_2O_8^{2-}$

phthalate, $C_8H_4O_4^{2-}$

polysulfide, S_x^{2-}

selenate, SeO_4^{2-}

sulfate, SO_4^{2-}

sulfite, SO_3^{2-}

tartrate, $C_4H_4O_6^{2-}$

tellurate, TeO_4^{2-}

tetraborate, $B_4O_7^{2-}$

thiosulfate, $S_2O_3^{2-}$

tungstate, WO_4^{2-}

zincate, ZnO_2^{2-}

3–

aluminate, AlO_3^{3-}

arsenate, AsO_4^{3-}

borate, BO_3^{3-}

citrate, $C_6H_5O_7^{3-}$

phosphate, PO_4^{3-}

4–

orthosilicate, SiO_4^{4-}

pyrophosphate, $P_2O_7^{4-}$

5–

tripolyphosphate, $P_3O_{10}^{5-}$

Cations

1+

ammonium, NH_4^+

hydronium, H_3O^+

Exercise 3–2: Name the following substances.

1. $FeSO_3$ _____

2. $Cu(NO_3)_2$ _____

3. Hg_2Cl_2 _____

4. $AgBr$ _____

5. $KClO_3$ _____

6. $MgCO_3$ _____

7. BaO_2 _____

8. KO_2 _____

9. SnO_2 _____

10. $Pb(OH)_2$ _____

11. $Ni_3(PO_4)_2$ _____

12. $CuCH_3COO$ _____

13. N_2O_4 _____

14. Rb_3P _____

15. S_8 _____

16. Fe_2O_3 _____

17. $(NH_4)_2SO_3$ _____

18. $Ca(MnO_4)_2$ _____

19. PF_5 _____

20. LiH _____

Exercise 3–3: Write formulas for the following substances.

1. vanadium(V) oxide

2. dihydrogen monoxide

3. ammonium oxalate

4. polonium(VI) thiocyanate

5. tetraphosphorus decaoxide

6. zinc hydroxide

7. potassium cyanide

8. cesium tartrate

9. oxygen molecule

10. mercury(II) acetate

11. silver chromate

12. tin(II) carbonate

13. sodium hydrogen carbonate

14. manganese(VII) oxide

15. copper(II) dihydrogen phosphate

16. francium dichromate

17. calcium carbide

18. mercury(I) nitrate

19. cerium(IV) benzoate

20. potassium hydrogen phthalate

Chapter 4
Ternary Nomenclature: Acids and Salts

The halogens, with their variable oxidation numbers, allow for a great variety of compounds. The problem arises on how these compounds should be named. For example, chlorine is found with a different oxidation state in each of the following five compounds:

$$HClO_4 \quad (Cl = 7+)$$

$$HClO_3 \quad (Cl = 5+)$$

$$HClO_2 \quad (Cl = 3+)$$

$$HClO \quad (Cl = 1+)$$

$$HCl \quad (Cl = 1-)$$

A good way to learn ternary nomenclature is to start with a certain "home base" polyatomic ion. This is the polyatomic ion ending with the suffix *–ate* (see page 16). Remembering that salts are named by adding the name of the metallic ion (cation) to the nonmetallic polyatomic ion (anion), the following rules apply:

Number of Oxygen Atoms (Compared to Home Base)	Polyatomic Ion Name		Acid Name (H⁺ Combined with Polyatomic Ion)	
Plus One Oxygen Atom	ClO_4^-	*perchlorate*	$HClO_4$	*perchloric* acid
Home Base	ClO_3^-	chlor*ate*	$HClO_3$	chlor*ic* acid
Minus One Oxygen Atom	ClO_2^-	chlor*ite*	$HClO_2$	chlor*ous* acid
Minus Two Oxygen Atoms	ClO^-	*hypo*chlor*ite*	$HClO$	*hypo*chlor*ous* acid
No Oxygen Atoms	Cl^-	chlor*ide*	HCl^*	*hydro*chlor*ic* acid

*Binary compounds containing hydrogen and a nonmetallic ion, such as hydrogen chloride, form acids when dissolved in water. The name of the resulting acid is derived by adding the prefix *hydro-* to the root name followed by the suffix *-ic* and the word acid. Thus, HCl gas is called hydrogen chloride (hydrogen monochloride), but is known as hydrochloric acid in aqueous solution.

Common Binary Acids

Formula	Name	Anion
HF(aq)	*hydro*fluor*ic* acid	F⁻, fluor*ide* ion
HCl(aq)	*hydro*chlor*ic* acid	Cl⁻, chlor*ide* ion
HBr(aq)	*hydro*brom*ic* acid	Br⁻, brom*ide* ion
HI(aq)	*hydro*iod*ic* acid	I⁻, iod*ide* ion
H_2S(aq)	*hydro*sulfur*ic* acid	S²⁻, sulf*ide ion*

Many common acids contain only oxygen, hydrogen, and a nonmetallic ion or a polyatomic ion. Such acids are called **oxyacids**. The suffixes *-ous* and *-ic* give the oxidation state of the atom bonded to the oxygen and the hydrogen. The *-ous* suffix always indicates the lower oxidation state and *-ic* the higher.

Common Oxyacids

Formula	Name	Anion	
$HClO_4$	*per*chlor*ic* acid	ClO_4^-	*per*chlor*ate*
$HClO_3$	chlor*ic* acid	ClO_3^-	chlor*ate*
$HClO_2$	chlor*ous* acid	ClO_2^-	chlor*ite*
$HClO$	*hypo*chlor*ous* acid	ClO^-	*hypo*chlor*ite*
HNO_3	nitr*ic* acid	NO_3^-	nitr*ate*
HNO_2	nitr*ous* acid	NO_2^-	nitr*ite*
H_2SO_4	sulfur*ic* acid	SO_4^{2-}	sulf*ate*
H_2SO_3	sulfur*ous* acid	SO_3^{2-}	sulf*ite*
CH_3COOH or $HC_2H_3O_2$	acet*ic* acid	CH_3COO^- or $C_2H_3O_2^-$	acet*ate*
H_2CO_3	carbon*ic* acid	CO_3^{2-}	carbon*ate*
$H_2C_2O_4$	oxal*ic* acid	$C_2O_4^{2-}$	oxal*ate*
H_3PO_4	phosphor*ic* acid	PO_4^{3-}	phosph*ate*

Exercise 4–1: Name the following compounds.

1. HIO_3

2. $NaBrO_2$

3. $Ca_3(PO_4)_2$

4. HIO_4

5. $Fe(IO_2)_3$

6. $HAt(aq)$

7. C_6H_5COOH

8. $Hg_2(IO)_2$

9. H_3PO_3

10. NH_4BrO_3

Exercise 4–2: Write formulas for the following compounds.

1. tartaric acid

2. calcium hypochlorite

3. hydrotelluric acid

4. copper(II) nitrite

5. carbonic acid

6. hypoiodous acid

7. cyanic acid

8. phthalic acid

9. tin(IV) chromate

10. selenic acid

DO YOU KNOW YOUR ACIDS?

-IC from *-ATE*   *-OUS* from *-ITE*   *HYDRO-, -IC, -IDE*

Exercise 4–3: Complete the following table.

Name of Acid	Formula of Acid	Name of Anion
*hydro*chlor*ic* acid	HCl	chlor*ide*
sulfur*ic* acid	H_2SO_4	sulf*ate*
	HI	
		sulf*ite*
chlor*ous* acid		
		nitr*ate*
	CH_3COOH or $HC_2H_3O_2$	
*hydro*brom*ic* acid		
		sulf*ide*
	HNO_2	
chrom*ic* acid		
		phosph*ate*

Notes

Chapter 5
Complex Ion Formulas and Nomenclature

Coordination Chemistry

Coordination compounds usually involve transition metal ions, which are typically colored and often paramagnetic. A neutral coordination compound normally consists of a *complex ion*—a transition metal ion with attached ligands—and *counterions*, which are anions or cations needed to produce a neutral compound.

- Complex ion — Transition metal ion with attached ligands
- Counterions — Anions or cations

Example

$[Co(NH_3)_6]Cl_3$ $[Co(NH_3)_6]^{3+}$ is the complex ion consisting of the transition metal ion Co^{3+} and six attached NH_3 ligands, and the Cl^-'s are the counterions.

Ligands are Lewis bases with lone pairs of electrons. These electrons can be donated to the empty *d* orbitals of a transition metal ion or aluminum ion, which in turn can act as a Lewis acid. This sharing of electrons in which one partner does all of the donating is called a *coordinate covalent bond*.

Common ligands include H_2O (aqua), NH_3 (ammine), Cl^- (chloro), CN^- (cyano), and OH^- (hydroxo).

Some ligands such as the carbonate ion, CO_3^{2-}, or ethylenediamine, $H_2NCH_2CH_2NH_2$ (may be abbreviated *en*), can provide more than one pair of electrons to a Lewis acid.

The number of ligands attached to the central metal ion is referred to as the *coordination number* of the metal. Two, four, and six are the most common coordination numbers. Often the number of ligands attached is two times the metal's oxidation number.

The colors of coordination compounds are quite varied. Transition metals with completely filled *d* sublevels tend to form white compounds and give colorless solutions.

Complex ions are typically formed by the reversible stepwise addition of ligands one at a time to the metal ion. The stepwise formation of each complex ion will have an associated formation constant (equilibrium constant). The greater the formation constant, the more stable the complex ion.

Naming Complex Ions

Complex cations, such as $[Cr(H_2O)_6]^{3+}$, are named by giving the number and name of the groups attached to the central metal atom followed by the name of the central atom, with its oxidation number indicated by a Roman numeral in parentheses. Note that the formula of the complex ion is written in brackets with the overall charge written outside the brackets.

e.g., hexaaquachromium(III)

Complex anions, such as $[PtCl_6]^{2-}$, are named by giving the number and name of the groups attached to the central metal atom, followed by the root name of the central atom, and ending with the suffix *-ate*. The oxidation number of the metal is again shown in parentheses.

e.g., hexachloroplatinate(IV)

23

IUPAC Rules for Naming Coordination Compounds

Note: IUPAC stands for the International Union of Pure and Applied Chemistry.

1. The cation is always named before the anion, with a space between the names.

2. In naming a complex ion, the ligands are named before the metallic ion.

3. The names of ligands are given special endings. Names of some common ligands are listed in the table below. Notice that an *-o* ending is used in place of any *-ide* ending for anions. For anions ending in *-ate*, the letter *-o* is substituted for the *-e*. For neutral ligands, the name of the molecule is used, with the exception of H_2O, NH_3, CO, and NO.

Names of Some Common Ligands

Anions

fluoro	F^-
chloro	Cl^-
bromo	Br^-
iodo	I^-
carbonato	CO_3^{2-}
cyano	CN^-
hydrido	H^-
hydroxo	OH^-
nitrato	NO_3^-
nitro	NO_2^-
oxalato	$C_2O_4^{2-}$ or *ox*
oxo	O^{2-}
sulfato	SO_4^{2-}
thiocyanato	SCN^-
thiosulfato	$S_2O_3^{2-}$

Neutral Ligands

aqua	H_2O
ammine*	NH_3
benzene	C_6H_6
carbonyl**	CO
ethylenediamine	$H_2NCH_2CH_2NH_2$ or *en*
methylamine	CH_3NH_2
dimethylamine	$(CH_3)_2NH$
nitrosyl	NO

*Note the spelling ammine (two m's) for a neutral ammonia ligand. An amine (one m) is a derivative of ammonia found in organic compounds.

**Ligand is attached via the carbon atom.

4. Ligands are named first, in alphabetical order, followed by the metal name. The Greek prefixes *di = 2, tri = 3, tetra = 4, penta = 5, hexa = 6* and so on are used to denote the number of simple ligands present. (*Mono* is omitted for one ligand.) For more complicated ligands the prefixes *bis* (twice), *tris* (thrice), *tetrakis* (four times), *pentakis* (five times), and *hexakis* (six times) are used. *Note:* Prefixes do not affect the alphabetical order.

5. Many metals exhibit variable oxidation numbers. The oxidation number of the metal is designated by a Roman numeral in parentheses following the name of the complex ion.

6. For anionic complex ions (those having negative charges), the suffix -*ate* is added to the root name of the metal. The English name for the metal is used, except in the following cases where the root Latin name is used to identify the metal (see table below). The suffix -*ate* indicates that the complex ion is an anion. Neutral coordination complexes and cationic complex ions do not use suffixes.

Latin Names for Some Metals Ions Found in Anionic Complex Ions	
copper	cuprate
gold	aurate
iron	ferrate
lead	plumbate
silver	argentate
tin	stannate

7. In the case of complex-ion isomerism, the names *cis* or *trans* may precede the formula or name of the complex ion to indicate the spatial arrangement of the ligands. *Cis* means the ligands occupy adjacent coordination positions; *trans* means opposite positions. In the examples given below, the trans isomers appear on the left and their cis isomers on the right.

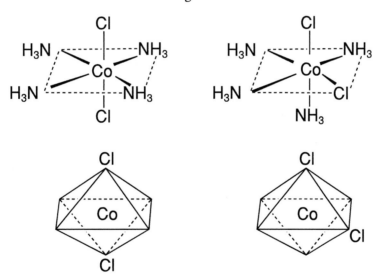

Determining oxidation states with a complex ion is fairly easy. For example, in the compound $Na[Al(OH)_4]$, we know the sodium ion must be 1+ and, therefore, the tetrahydroxoaluminate complex ion must be 1–. The 1– results because the Al is 3+ and the hydroxide is 4 times 1–; (3+) + (4–) equals 1–. In the compound $[Cr(H_2O)_4Cl_2]Cl$ we know the charge of the chloride counterion is 1–. Therefore, the complex ion tetraaquodichlorochromium(III) must have a total charge of 1+. The charge on four water molecules is zero (a neutral compound), and the two chloride ions equal 2 times 1– or 2–. The charge on the chromium must be 3+ because [3+ plus 4(0) plus 2(1–)] = 1+, the charge of the complex cation.

Examples

$[Cr(H_2O)_4Cl_2]^+$ tetraaquadichlorochromium(III) ion

Explanation: This cation's name alphabetically places the ligand *aqua* before *chloro*. The prefixes *tetra* and *di* indicate the number of each ligand attached to the central chromium ion. The overall charge of the complex ion is 1+ because Cr is 3+, the water molecules are neutral, and the two chloro ligands are each 1–.

$[Cr(H_2O)_4Cl_2]NO_3$ tetraaquadichlorochromium(III) nitrate

Explanation: This coordination compound uses the complex cation name first followed by the anion name, nitrate. Notice that the charge of the complex ion (1+) balances with the charge of the nitrate ion (1–), so no additional subscripts are required.

$[Cr(H_2O)_2Cl_4]^-$ diaquatetrachlorochromate(III) ion

Explanation: This anion's name alphabetically places the ligand *aqua* before *chloro*. The prefixes *di* and *tetra* indicate the number of each ligand attached to the central chromium ion. The overall charge of the ion is 1– because Cr is 3+ and the four chloro ligands are each 1–. Note that the name indicates that this is an anion because of the *–ate* ending.

$Mg[Cr(H_2O)_2Cl_4]_2$ magnesium diaquatetrachlorochromate(III)

Explanation: This coordination compound uses magnesium as the cation name first followed by the complex anion's name. Notice that the charge of the complex ion (1–) does not balance with the charge of the magnesium ion (2+), so a subscript of 2 was required for the complex ion.

$[Pt(NH_3)_4]\,[PtCl_6]$ tetraammineplatinum(II) hexachloroplatinate(IV)

Explanation: This coordination compound is made up of a complex cation and a complex anion. The complex cation has an overall charge of 2+ due to the platinum ion, since ammonia is neutral. The complex anion has an overall charge of 2– with the platinum having an oxidation state of 4+ and the six chloro ligands each having an oxidation state of 1–. Notice that the cation name ends with the metal name and a Roman numeral for the metal oxidation state (2+) whereas the anion name ends with the ending *–ate* and a Roman numeral for the metal oxidation state (4+).

$K_4[Ni(CN)_2(ox)_2]$ potassium dicyanobis(oxalato)nickelate(II)

Explanation: This coordination compound uses potassium as the cation name first followed by the complex anion's name. Notice that the charge of the complex ion (4–) does not balance with the charge of the potassium ion (1+), so a subscript of 4 was required for the potassium ion. The anion contains both a simple ligand, *cyano*, with a prefix of *di* indicating 2, as well as a more complicated ligand, *oxalato*, which has a prefix of *bis* indicating 2 and an abbreviation of *(ox)*.

Exercise 5–1: Name the following.

1. $[Ag(NH_3)_2]Cl$ _____

2. $(NH_4)_3[Fe(SCN)_6]$ _____

3. $Na_2[Ni(CN)_4]$ _____

4. $[Fe(ox)_3]^{3-}$ _____

5. $[Co(NH_3)_6]Br_2$ _____

6. $[Cr(H_2O)_4Cl_2]Cl$ _____

7. $[Pt(NH_3)_2]Cl_2$ _____

8. $K_2[Cu(CN)_4]$ _____

9. $[Cr(H_2O)_6](NO_3)_3$ _____

10. $[Co(en)_3]Br_3$ _____

Exercise 5–2: Write formulas for the following.

1. potassium hexacyanoferrate(III)

   _____

2. sodium hexafluoroaluminate

   _____

3. diamminesilver ion

   _____

4. tetraamminezinc nitrate

   _____

5. sodium tetrahydroxochromate(III)

   _____

6. _trans_-dichlorobis(ethylenediamine)cobalt(III) chloride

   _____

7. hexaammineruthenium(III) tetrachloronickelate(II)

   _____

8. tetraamminecopper(II) pentacyanohydroxoferrate(III)

   _____

9. sodium tetracyanocadmate

   _____

10. diamminezinc iodide

   _____

Chapter 6
Organic Nomenclature and Simple Reactions

Organic chemistry involves the chemistry of compounds containing carbon. Because of carbon's unique ability to bond with itself in a large variety of ways, the field of organic chemistry involves more than nine million compounds. Not only are organic compounds a vital constituent of all living things but they also contribute to the wide variety of materials we use in our modern lifestyles. Plastics we recycle, gasoline that we burn, and synthetic materials that we wear are all organic. In addition to carbon, organic compounds contain other elements as well, most commonly hydrogen, oxygen, nitrogen, sulfur, phosphorus, and the halogens.

The properties of organic compounds differ from those of inorganic compounds most notably in their relatively low melting points (generally under 300 °C), low boiling points, solubility in other organic solvents rather than water, and poor electrical conductivity. Organic reactions often have slow reaction rates and provide low yields of product due to the tendency for many side reactions to occur.

References to organic compounds may be made either by name or by formula. Because of the variations in the structures of organic compounds, three different types of formulas are commonly used. The definition and an example of each type of formula are given below, using ethanol (ethyl alcohol) as an example.

Molecular formulas indicate the type and number of each atom in the compound but give no information about the bonds or the structure of the compound.

$$C_2H_5OH$$

Structural formulas indicate the bonding arrangements of all atoms in the structure, showing all of the bonds present.

$$
\begin{array}{cc}
\text{H} & \text{H} \\
| & | \\
\text{H---C---C---O---H} \\
| & | \\
\text{H} & \text{H}
\end{array}
$$

Condensed structural formulas are shorthand representations that leave the bond lines out, yet still indicate what is bonded to each carbon or other atom.

$$CH_3CH_2OH$$

Hydrocarbons

A large number of organic compounds fall into a category known as the hydrocarbons. As the name implies, hydrocarbons consist of only carbon and hydrogen but even those two elements can produce a wide variety of compounds. Hydrocarbons can also be subdivided into two groups—open chain or acyclic hydrocarbons, and closed or cyclic (ring) hydrocarbons.

An acyclic hydrocarbon consists of a chain of carbon atoms (e.g., C—C—C—C) with hydrogen atoms attached to the carbon atoms along the periphery of the chain. Since the carbon atoms in each chain can be covalently bonded with either single, double or triple pairs of shared electrons, three groups of acyclic hydrocarbons emerge—alkanes, alkenes, and alkynes.

Alkanes are hydrocarbons in which there are only single covalent bonds between the carbon atoms. The general formula for an alkane is C_nH_{2n+2}, where n is the number of carbon atoms in the chain. Because carbon exhibits only single bonds and it has four effective electron pairs, it displays sp^3 hybridization in all alkanes. The length of the carbon chain tends to affect physical properties such as boiling point and melting point due to the variation in strength of the dispersion forces. Naming alkanes is straightforward— simply use the root name or stem listed in the box below to indicate the number of carbons in the chain and add the suffix *–ane* to the end. Thus an alkane containing six carbons (C_6H_{14}) is named hexane, while one containing three carbons (C_3H_8) is named propane.

Root Names or Stems for Organic Compounds

1	meth-	6	hex-
2	eth-	7	hept-
3	prop-	8	oct-
4	but-	9	non-
5	pent-	10	dec-

Alkenes are hydrocarbons in which there is at least one double covalent bond between two carbon atoms. The general formula for an alkene with one double bond is C_nH_{2n}, where n is the number of carbon atoms in the chain. Each doubly-bonded carbon atom has one double bond and two single bonds, giving it three effective electron pairs and sp^2 hybridization. To name alkenes, indicate where the double bond occurs by numbering the carbon chain starting with the end closest to the double bond. Then use the root name or stem to indicate the number of carbons in the chain and add the suffix *–ene* to the end. For example,

$$\begin{array}{ccccc} & H & H & H & H \\ & | & | & | & | \\ H- & C= & C- & C- & C-H \\ & & & | & | \\ & & & H & H \end{array}$$

is named 1–butene

$$\begin{array}{ccccc} & H & H & H & H \\ & | & | & | & | \\ H- & C- & C= & C- & C-H \\ & | & & & | \\ & H & & & H \end{array}$$

is named 2–butene

For chains that contain two double bonds, list the location of the two double bonds and use the suffix *–diene* at the end.

$$\begin{array}{ccccc} & H & H & H & H \\ & | & | & | & | \\ H- & C= & C- & C= & C-H \end{array}$$

is named 1,3–butadiene

Alkynes are hydrocarbons in which there is at least one triple covalent bond between two carbon atoms. The general formula for an alkyne is C_nH_{2n-2}, where n is the number of carbon atoms in the chain. Each triply-bonded carbon atom has one triple bond and one single bond, giving it two effective electron pairs and sp hybridization. To name alkynes, indicate where the triple bond occurs by numbering the carbon chain starting with the end closest to the triple bond. Then use the root name or stem to indicate the number of carbons in the chain and add the suffix *–yne* to the end.

The most common cyclic organic compounds are derivatives of benzene, C_6H_6. This six-member carbon ring has three alternating double bonds and can be found with many different components attached to it, including other benzene rings. Other cyclic hydrocarbons include cyclobutane and cyclopentane.

Naming Branched Hydrocarbons

Thus far, the naming of different types of straight-chain hydrocarbons has been considered. Hydrocarbons can also be formed by attaching hydrocarbon groups onto a hydrocarbon chain. These are generally referred to as branched hydrocarbons. General rules for naming these compounds are as follows:

- For *alkanes*, locate the longest continuous chain of carbon atoms and count the number of atoms—this chain determines the root name or stem for the compound. For *alkenes and alkynes*, determine the number of carbon atoms in the longest carbon chain that contains the multiple bond—this will give the root name or stem for the compound.

- Look for any groups other than hydrogen that may appear in the molecule. These are called substituent groups. There will be a special prefix for each such substituent group. For example:

Substituent Group		Prefix
$-CH_3$	1-carbon group	methyl
$-CH_2CH_3$	2-carbon group	ethyl
$-CH_2CH_2CH_3$	3-carbon group	propyl
$-Cl$	chlorine atom	chloro
$-Br$	bromine atom	bromo

Note: Take the time to memorize the first 10 prefixes—it will help in naming thousands of compounds.

- If more than one substituent group of any kind is present, use Greek prefixes to indicate the number. For example, use:

di-	for 2
tri-	for 3
tetra-	for 4

- Number the longest continuous carbon chain beginning with the end of the chain nearest the substituent. Use the numbers to designate the location of the substituent group(s). If there are more than two substituent groups of the same kind, give each group a number.

- *For alkanes*, always count from the end of the chain that will give the lowest possible combination of numbers. In other words, start counting from the end closest to the substituent group. *For alkenes and alkynes*, always count from the end of the chain that the multiple bond is closer to; disregard the branch chain rule used for alkanes.

- Use hyphens to separate numbers from names, and commas to separate numbers from each other. Substituent groups should be listed alphabetically, disregarding the Greek prefixes.

  Example 1

  $$CH_3CH_2CH(CH_3)CH(CH_3)CH_3 \qquad \textit{Condensed Formula}$$

  $$\overset{5}{C}H_3 - \overset{4}{C}H_2 - \overset{3}{C}H - \overset{2}{C}H - \overset{1}{C}H_3 \qquad \textit{Structural Formula}$$
  $$\qquad\qquad\quad | \quad\; |$$
  $$\qquad\qquad\quad CH_3 \; CH_3$$

  Name: 2,3-dimethylpentane

Example 2

$$CH_3CH(CH_3)C(CH_3)(CH_2CH_3)CH_2CH_2CH_3$$ *Condensed Formula*

$$\underset{1}{CH_3}-\underset{2}{\overset{\displaystyle CH_3}{\underset{|}{CH}}}-\underset{3}{\overset{\displaystyle CH_3}{\underset{\underset{CH_2CH_3}{|}}{\overset{|}{C}}}}-\underset{4}{CH_2}-\underset{5}{CH_2}-\underset{6}{CH_3}$$ *Structural Formula*

Name: 3-ethyl-2,3-dimethylhexane

Other Organic Functional Groups

Many remaining organic compounds can be divided into groups based on their unique structure. They include variations with oxygen and nitrogen and the means by which they bond to carbon and hydrogen. Each group has recognizable structural characteristics and names which are an extension of the general hydrocarbon naming rules.

Notes: R = a general symbol to represent any carbon chain or ring.
The systematic name is the official IUPAC name for the compound.

A. *Alcohols*   **R—OH**

1. Can be oxidized to aldehydes, carboxylic acids or ketones.

2. Systematic name:

   • Indicate the number of the carbon to which the —OH group is attached.

   • Drop the "e" from the name of the corresponding alkane, and add the suffix *-ol*.

B. *Ethers*   **R—O—R′**

1. Relatively unreactive compounds.

2. Systematic name:

   • Use the root for R, add the suffix *-oxy* followed by the corresponding alkane for R′.

C. *Carboxylic Acids*   **R—COOH** $R-\overset{\displaystyle \overset{O}{||}}{C}-O-H$

1. Formed from the oxidation of primary alcohols (—OH on the end carbon).

2. Tend to be polar and form hydrogen bonds.

3. Systematic name:

   • Drop the "e" from the name of the corresponding alkane and add the suffix *-oic* followed by the word *acid*.

D. *Esters*   R—COO—R′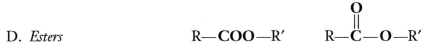

1. Derived from an acid and an alcohol.

2. Low molecular mass esters tend to have fruity odors.

3. Systematic name:

   • Name the alcohol part of the compound (the group to the right of the **COO**) with its *-yl* ending.

   • Name the acid part of the compound (the group to the left of the **COO** and including the carbon in **COO**) with an *-oate* ending.

E. *Aldehydes*   R—CHO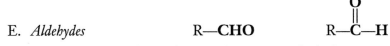

1. Derived from the oxidation of a primary alcohol.

2. Can be oxidized to a carboxylic acid.

3. Systematic name:

   • Take the parent alkane name, drop the "e", and add the suffix *-al*.

F. *Ketones*   R—CO—R′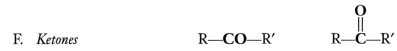

1. Derived from the oxidation of a secondary alcohol.

2. Systematic name:

   • Consider the total number of carbons—take the parent alkane name, drop the "e", and add the suffix *-one*.

G. *Amines*   R—NH$_2$ R—N—H

1. Are weakly basic compounds (may be considered to be derivatives of ammonia).

2. Tend to have offensive odors (think of decaying fish).

3. Systematic name:

   • Name the R group and add the word *amine*.

H. *Amides*   R—CO—NH$_2$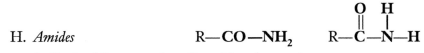

1. Derived from a carboxylic acid and an amine.

2. Significant in synthetic polymers and proteins.

3. Systematic name:

   • Consider the total number of carbons—take the parent alkane name, drop the "e", add the suffix *-yl*, and add the word *amide*.

Organic Functional Groups			
Category	**Functional Group**	**General Formula**	**Example**
Alcohols	$-OH$	$R-OH$	CH_3OH methanol
Ethers	$-O-$	$R-O-R'$	CH_3OCH_3 methoxymethane
Carboxylic acids	$\overset{\displaystyle O}{\overset{\|}{-C-O-H}}$	$\overset{\displaystyle O}{\overset{\|}{R-C-O-H}}$	CH_3COOH ethanoic acid
Esters	$\overset{\displaystyle O}{\overset{\|}{-C-O-}}$	$\overset{\displaystyle O}{\overset{\|}{R-C-O-R'}}$	CH_3COOCH_3 methylethanoate
Aldehydes	$\overset{\displaystyle O}{\overset{\|}{-C-H}}$	$\overset{\displaystyle O}{\overset{\|}{R-C-H}}$	CH_3CHO ethanal
Ketones	$\overset{\displaystyle O}{\overset{\|}{-C-}}$	$\overset{\displaystyle O}{\overset{\|}{R-C-R'}}$	$CH_3CH_2COCH_3$ 2-butanone
Amine	$-NH_2$	$R-NH_2$	CH_3NH_2 methylamine
Amide	$\overset{\displaystyle O}{\overset{\|}{-C-NH_2}}$	$\overset{\displaystyle O}{\overset{\|}{R-C-NH_2}}$	$CH_3CH_2CONH_2$ propylamide

Exercise 6–1: Using condensed formulas provided, name the following hydrocarbon compounds. It may be helpful to draw the structural formula first.

1. $CH_3CH_2CH_2CH(CH_3)CH_3$

2. $CH_3CH(C_2H_5)CH_2CH_3$

3. $CH_3CH_2CHClCH_2CH_2CH_2CH_2Cl$

4. $CH_3CH_2CH_2CH_2OH$

5. $CH_3CH_2CH_2CH_2COOH$

6. $CH_2=C(CH_3)CH_2CH(CH_3)CH_3$

7. $(CH_3)_2CHCl$

8. $CH_3C(CH_3)_2CH_2C(CH_3)_2CH_2CH_2CH_3$

9. $CH_3C(CH_3)_2CH=C(CH_3)CH_2CH_3$

10. $CH_3C{\equiv}CCH_3$

Exercise 6-2: Draw the condensed formula for each of the following compounds.

1. 2,3-dimethyl-2-butene

2. 4-ethyl-2-hexyne

3. 3,3,6-trimethylnonane

4. 3-ethyl-4-propylheptane

5. 3-octanol

6. 2-methyl-2-pentene

7. 5-methyl-1-hexene

8. 2,2,4,5-tetramethylhexane

9. propanoic acid

10. 2-pentyne

Simple Organic Reactions

Although organic compounds can undergo a great many reactions, some of the most common types of reactions include addition, substitution, combustion, and esterification.

Addition reactions generally occur when halogens (halogenation) or hydrogens (hydrogenation) are added to alkenes or alkynes. The net effect is that the double or triple bond is eliminated and the product contains singly-bonded components. This process takes an unsaturated compound and causes it to become saturated. For example,

Bromine added to ethene forms 1,2-dibromoethane.

$$Br_2 + CH_2{=}CH_2 \longrightarrow CH_2Br{-}CH_2Br$$

Hydrogen added to ethene (at high temperature and with a metal catalyst) forms ethane.

$$H_2 + CH_2{=}CH_2 \xrightarrow{Pt} CH_3{-}CH_3$$

Chlorine added to ethyne forms 1,2-dichloroethene. Excess chlorine will result in 1,1,2,2-tetrachloroethane.

$$Cl_2 + CH{\equiv}CH \longrightarrow CHCl{=}HCl + Cl_2 \longrightarrow CHCl_2{-}CHCl_2$$

Substitution reactions occur when an atom attached to a carbon is removed and something else takes its place. No change in bonding occurs. A common substitution is when a halogen is added to a saturated (only single covalent bonds present) hydrocarbon compound. For example,

Chlorine added to methane forms chloromethane. Often this compound as well as some dichloromethane, trichloromethane, and tetrachloromethane will be formed.

$$Cl_2 + CH_4 \longrightarrow CH_3Cl + HCl$$

Combustion reactions involve the oxidation of an organic compound. Incomplete combustion products may include carbon, carbon monoxide, and water. When an organic compound is burned completely in air, the products of complete combustion are carbon dioxide and water.

Ethane burns in oxygen.

$$2C_2H_6 + 7O_2 \longrightarrow 4CO_2 + 6H_2O$$

Ethanol is burned in oxygen gas.

$$C_2H_5OH + 3O_2 \longrightarrow 2CO_2 + 3H_2O$$

Esterification reactions occur when a carboxylic acid is combined with an alcohol. The products which form are an ester and water. The reactions are usually aided by the addition of a small amount of an inorganic acid such as sulfuric acid that acts as a catalyst or dehydrating agent. In essence the hydroxyl group from the alcohol and the acidic hydrogen from the carboxylic acid are removed to form water. The ester group bonds to the alcohol's remaining carbon chain. For example, ethanoic acid (acetic acid) when combined with methanol (methyl alcohol) will form methyl acetate and water.

$$CH_3COOH + CH_3\mathbf{OH} \xrightarrow{H^+} CH_3COOCH_3 + \mathbf{HOH}$$

Ethanoic (acetic) acid is reacted with 1-octanol in the presence of sulfuric acid catalyst.

$$CH_3COOH + C_8H_{17}\mathbf{OH} \xrightarrow{H^+} CH_3COOC_8H_{17} + \mathbf{HOH}$$

Exercise 6–3: Predict and balance the following organic reactions.

1. Ethanol (ethyl alcohol) is burned completely in air.

2. Propane gas is heated with chlorine gas.

3. Ethanol (ethyl alcohol) and methanoic acid (formic acid) are mixed and warmed.

4. Ethene gas is bubbled through a solution of bromine.

5. Hydrogen gas is added to 2-pentene in the presence of a metal catalyst.

6. Octane is burned in oxygen.

7. 2-Butene is combined with hydrogen gas in the presence of a nickel catalyst.

8. Ethanoic acid is combined with propanol.

9. An excess of chlorine gas is added to pure ethyne (acetylene) gas.

10. A limited amount of liquid bromine is added to an excess of benzene (C_6H_6).

Some possible questions for organic reactions:

- Draw a Lewis structure or identify molecular geometry.
- Identify the hybridization of an atom.
- Count the number of sigma and/or pi bonds.
- Use mole ratios to determine the total moles of products formed.

Example   Butane is burned in oxygen.

If 0.40 moles of butane is burned, how many total moles of products will be made?

Solution $2C_4H_{10} + 13O_2 \longrightarrow 8CO_2 + 10H_2O$

            3.6 moles of products

Example   Ethyne is burned in oxygen.

Draw a Lewis structure of ethyne.

Solution $2C_2H_2 + 5O_2 \longrightarrow 4CO_2 + 2H_2O$

            $H-C\equiv C-H$

Example   Ethanoic acid is reacted with 1-propanol in the presence of sulfuric acid.

List the number of sigma and pi bonds in an ethanoic acid molecule.

Solution $CH_3COOH + C_3H_5OH \longrightarrow CH_3COOC_3H_5 + H_2O$

            7 sigma bonds and 1 pi bond

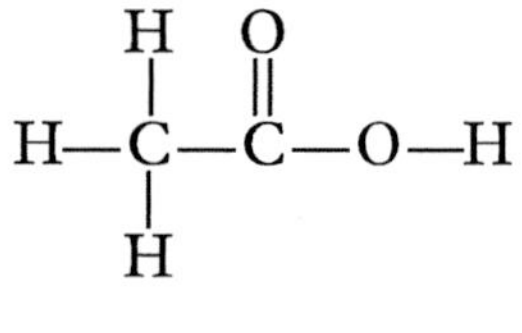

$$\begin{array}{ccc} & H & O \\ & | & \| \\ H- & C-C & -O-H \\ & | & \\ & H & \end{array}$$

Example   Hexane is burned in oxygen.

Identify the hybridization of carbon on both the reactant and product side.

Solution $2C_6H_{14} + 19O_2 \longrightarrow 12CO_2 + 14H_2O$

            sp³ reactant and *sp* product

Chapter 7
Balancing Molecular Equations

Chemists write chemical equations to illustrate what is happening during a chemical reaction—bonds are broken, atoms are rearranged, and new bonds are formed. Every chemical reaction supports the law of conservation of matter. This means that in every reaction, the number of atoms of each type of element contained within the reactants must be the same as the number of atoms of each type of element contained within the products.

Balancing equations is a process which assures that equations are written properly to support the law of conservation of matter. An equation cannot be balanced, however, until each reactant and product formula is written correctly. It is important to properly write the seven elements that are diatomic in their elemental form and also to use subscripts and parentheses appropriately when considering the oxidation numbers of ions. All compounds must be made neutral before beginning to balance atoms.

Balancing is accomplished by adding coefficients that multiply the number of atoms represented by the formula. For example, a coefficient of 2 in front of oxygen (e.g., $2O_2$) in an equation means that 4 oxygen atoms are represented. Unlike algebra, in chemistry a coefficient does not need to be outside parentheses or brackets to be distributed. A coefficient applies to the complete substance; however, it no longer applies when a plus sign (+) or arrow ($\longrightarrow$) is encountered. For example,

$3(NH_4)_2CO_3$ shows 6 nitrogen, 24 hydrogen, 3 carbon, and 9 oxygen atoms

$3MgCl_2$ + NaBr indicates 3 magnesium, 6 chlorine, 1 sodium, and 1 bromine atoms

Tips for Balancing Equations

- Be sure each molecular formula is written correctly and each compound is neutral.

- Mentally count or tally how many of each type of atom is present on each side of the equation.

- Begin by balancing elements that are found in only one substance on each side of the equation.

- Balance oxygen and hydrogen *last*—they usually balance out at the end or are easily balanced at the end by adjusting only the number of water molecules.

- If there is an odd number of atoms for an element on one side of the equation and an even number on the other side, the odd number will need to be evened out—so use a coefficient of 2 for that substance.

- If there are polyatomic ions that remain together as a unit during the reaction, count the polyatomic ion as a unit.

- When tallying atoms in an equation, be sure to adjust the count for each and every element that an added coefficient affects.

- Combustion reactions that don't seem to balance will often come out better if a coefficient of 2 is used for the hydrocarbon.

Some Examples

Hydrogen and chlorine gases react to form hydrogen chloride gas.

$$H_2 + Cl_2 \longrightarrow HCl$$

H	2	H	1
Cl	2	Cl	1

Because one hydrogen and one chlorine atom appear to have been lost on the product side, a coefficient of 2 is added to the HCl.

$$H_2 + Cl_2 \longrightarrow 2HCl$$

H	2	H	2
Cl	2	Cl	2

Potassium chlorate is heated and decomposes into potassium chloride and oxygen gas.

$$KClO_3 \xrightarrow{\Delta} KCl + O_2$$

K	1	K	1
Cl	1	Cl	1
O	3	O	2

Because there is an odd number of oxygen atoms on the left and an even number on the right, a coefficient of 2 is added to the potassium chlorate.

$$2KClO_3 \xrightarrow{\Delta} KCl + O_2$$

K	2	K	1
Cl	2	Cl	1
O	6	O	2

The oxygen atoms can now be balanced using a coefficient of 3; the potassium and chlorine atoms can be balanced using a coefficient of 2 in front of KCl.

$$2KClO_3 \xrightarrow{\Delta} 2KCl + 3O_2$$

K	2	K	2
Cl	2	Cl	2
O	6	O	6

Ammonium nitrate reacts with calcium phosphate to form ammonium phosphate and calcium nitrate.

$$NH_4NO_3 + Ca_3(PO_4)_2 \longrightarrow (NH_4)_3PO_4 + Ca(NO_3)_2$$

Balancing this equation can be simplified greatly by noticing that each polyatomic ion (both nitrate and phosphate ions) remains together as a group throughout the reaction. Note that the tallying below indicates this.

NH_4	1	NH_4	3
NO_3	1	NO_3	2
Ca	3	Ca	1
PO_4	2	PO_4	1

Beginning with calcium, add a coefficent of 3 on the right to balance the Ca. This also changes the nitrate count to 6 on the right, so adjust that by adding a coefficient of 6 on the left side.

$6NH_4NO_3 + Ca_3(PO_4)_2 \longrightarrow (NH_4)_3PO_4 + 3Ca(NO_3)_2$

NH_4	6	NH_4	3
NO_3	6	NO_3	6
Ca	3	Ca	3
PO_4	2	PO_4	1

Finally, a coefficient of 2 is needed in front of ammonium phosphate to complete the balancing.

$6NH_4NO_3 + Ca_3(PO_4)_2 \longrightarrow 2(NH_4)_3PO_4 + 3Ca(NO_3)_2$

NH_4	6	NH_4	6
NO_3	6	NO_3	6
Ca	3	Ca	3
PO_4	2	PO_4	2

As a double check, it is always wise to count the total number of each type of element indicated. For this reaction there is a total of 12 N, 24 H, 26 O, 3 Ca and 2 P atoms on each side.

Now it is time to apply the rules for balancing equations to some chemical reactions. It turns out that many chemical reactions can be organized into various groups—that is the focus of the sections that follow.

Exercise 7–1: Balance the following equations by adding coefficients as needed. Some equations may already be balanced.

1. ___Ca + ___HOH $\longrightarrow$ ___Ca(OH)$_2$ + ___H$_2$

2. ___Cl$_2$O$_7$ + ___H$_2$O $\longrightarrow$ ___HClO$_4$

3. ___Fe + ___O$_2$ $\longrightarrow$ ___Fe$_3$O$_4$

4. ___C$_6$H$_{14}$ + ___O$_2$ $\longrightarrow$ ___CO$_2$ + ___H$_2$O

5. ___Ca$_3$(PO$_4$)$_2$ + ___H$_2$SO$_4$ $\longrightarrow$ ___Ca(H$_2$PO$_4$)$_2$ + ___CaSO$_4$

6. ___AlCl$_3$ + ___AgNO$_3$ $\longrightarrow$ ___Al(NO$_3$)$_3$ + ___AgCl

7. ___HCl + ___CaCO$_3$ $\longrightarrow$ ___CO$_2$ + ___HOH + ___CaCl$_2$

8. ___WO$_3$ + ___H$_2$ $\longrightarrow$ ___W + ___H$_2$O

9. ___Cl$_2$ + ___H$_2$O $\longrightarrow$ ___HCl + ___HClO

10. ___Cl$_2$ + ___NaI $\longrightarrow$ ___NaCl + ___I$_2$

Synthesis Reactions occur when two or more reactants combine to form a single product. There are several common types of synthesis reactions.

A metal combines with a nonmetal to form a binary salt.

Example   A piece of lithium metal is dropped into a container of nitrogen gas.

$$6Li + N_2 \longrightarrow 2Li_3N$$

Metallic oxides and water form bases (metallic hydroxides).

Example   Solid sodium oxide is added to water.

$$Na_2O + HOH \longrightarrow 2NaOH$$

Solid magnesium oxide is added to water.

$$MgO + 2HOH \longrightarrow Mg(OH)_2$$

Nonmetallic oxides and water form acids. The nonmetal retains its oxidation number.

Example   Carbon dioxide is bubbled into water.

$$CO_2 + H_2O \longrightarrow H_2CO_3 \qquad \text{(Oxidation number of C is +4)}$$

Dinitrogen pentoxide is bubbled into water.

$$N_2O_5 + H_2O \longrightarrow 2HNO_3 \qquad \text{(Oxidation number of N is +5)}$$

Metallic oxides and nonmetallic oxides form salts.

Example   Solid sodium oxide is added to carbon dioxide.

$$Na_2O + CO_2 \longrightarrow Na_2CO_3$$

Solid calcium oxide is added to sulfur trioxide.

$$CaO + SO_3 \longrightarrow CaSO_4$$

Decomposition Reactions occur when a single reactant is broken down into two or more products.

Metallic carbonates decompose into metallic oxides and carbon dioxide.

Example   A sample of magnesium carbonate is heated.

$$MgCO_3 \xrightarrow{\Delta} MgO + CO_2$$

Metallic chlorates decompose into metallic chlorides and oxygen.

Example   A sample of magnesium chlorate is heated.

$$Mg(ClO_3)_2 \xrightarrow{\Delta} MgCl_2 + 3O_2$$

Ammonium carbonate decomposes into ammonia, water, and carbon dioxide.

Example   A sample of ammonium carbonate is heated.

$$(NH_4)_2CO_3 \xrightarrow{\Delta} 2NH_3 + H_2O + CO_2$$

Sulfurous acid decomposes into sulfur dioxide and water.

Example   A sample of sulfurous acid is heated.

$$H_2SO_3 \xrightarrow{\Delta} H_2O + SO_2$$

Carbonic acid decomposes into carbon dioxide and water.

Example   A sample of carbonic acid is heated.

$$H_2CO_3 \xrightarrow{\Delta} H_2O + CO_2$$

A binary compound may break down to produce two elements.

Example   Molten sodium chloride is electrolyzed.

$$2NaCl \longrightarrow 2Na + Cl_2$$

Hydrogen peroxide decomposes into water and oxygen.

Example $2H_2O_2 \longrightarrow 2H_2O + O_2$

Ammonium hydroxide decomposes into ammonia and water.

Example $NH_4OH \longrightarrow NH_3 + HOH$

Exercise 7–2: Predict and balance the following synthesis and decomposition reactions. Use abbreviations to indicate the phase of reactants and products where possible [i.e., (aq) (s) (l) (g)].

1. A sample of calcium carbonate is heated.

2. Sulfur dioxide gas is bubbled through water.

3. Solid potassium oxide is added to a container of carbon dioxide gas.

4. Liquid hydrogen peroxide is warmed.

5. Solid lithium oxide is added to water.

6. Molten aluminum chloride is electrolyzed.

7. A pea-sized piece of sodium is added to a container of iodine vapor.

8. A sample of carbonic acid is heated.

9. A sample of potassium chlorate is heated.

10. Solid magnesium oxide is added to sulfur trioxide gas.

Some possible questions for synthesis and decomposition reactions:

- Identify evidence of a chemical change.
- Predict the result when testing for gases with a burning splint.
- Predict a change in pH or acid–base indicator color.
- Use mole ratios to determine the total moles of products formed.

Example   Solid strontium carbonate is heated.

If the reaction is done in a test tube and a burning splint is inserted, what will be observed?

Solution $SrCO_3 \xrightarrow{\Delta} SrO + CO_2$

Carbon dioxide gas will extinguish the flame.

Example   A sample of zinc chlorate is heated.

If the reaction is done in a test tube and a burning splint is inserted, what will be observed?

Solution $Zn(ClO_3)_2 \xrightarrow{\Delta} ZnCl_2 + 3O_2$

Oxygen gas will cause the flame to burn brighter.

Example   Water is added to dinitrogen trioxide gas.

When this reaction occurs, will the pH increase, decrease or remain the same? Explain.

Solution $H_2O + N_2O_3 \longrightarrow 2HNO_2$

The pH will decrease (become more acidic) because nitrous acid is produced.

Example   Solid barium oxide is added to water.

If the water had several drops of bromthymol blue in it before the barium oxide was added, what color change would be observed as the reaction occurs? Explain.

Solution $BaO + H_2O \longrightarrow Ba(OH)_2$

$(BaO + H_2O \longrightarrow Ba^{2+} + 2OH^-$ net ionic equation)

The color of bromthymol blue will be green in the neutral water and then turn blue as the base, barium hydroxide, is produced.

Chapter 8
Single Replacement Reactions

Single replacement reactions are reactions that involve an element replacing one part of a compound. The products include the displaced element and a new compound. An element can only replace another element that is less active than itself.

General activity series for metals		
(most active)	Li Ca Na Mg Al Zn Fe Pb [H$_2$] Cu Ag Pt Au	(least active)

General activity series for nonmetals		
(most active)	F$_2$ Cl$_2$ Br$_2$ I$_2$	(least active)

Here are some common types of single replacement reactions.

Active metals replace less active metals from their compounds in aqueous solution.

Example   Magnesium turnings are added to a solution of iron(III) chloride.

$$3Mg + 2FeCl_3 \longrightarrow 2Fe + 3MgCl_2$$

Active metals replace hydrogen in water.

Example   Sodium is added to water.

$$2Na + 2HOH \longrightarrow H_2 + 2NaOH$$

Note: Water can be written as either HOH or H$_2$O, but writing it as HOH makes it more apparent that the first H is replaced in the reaction above.

Active metals replace hydrogen in acids.

Example   Lithium is added to hydrochloric acid.

$$2Li + 2HCl \longrightarrow H_2 + 2LiCl$$

Active nonmetals replace less active nonmetals from their compounds in aqueous solution.

Example   Chlorine gas is bubbled into a solution of potassium iodide.

$$Cl_2 + 2KI \longrightarrow I_2 + 2KCl$$

If a less reactive element is combined with a more reactive element in compound form, there will be no resulting reaction.

Example   Chlorine gas is bubbled into a solution of potassium fluoride.

$$Cl_2 + KF \longrightarrow \text{no reaction}$$

Example   Zinc is added to a solution of sodium chloride.

$$Zn + NaCl \longrightarrow \text{no reaction}$$

Exercise 8–1: Using the activity series, predict and balance the following single replacement reactions. Use abbreviations to indicate the appropriate phase of reactants and products where possible. *Note:* Not all of the reactions will occur. For those that do not, write "no reaction."

1. A piece of copper is dropped into a container of water.

2. Liquid bromine is added to a container of sodium iodide crystals.

3. An aluminum strip is immersed in a solution of silver nitrate.

4. Zinc pellets are added to a sulfuric acid solution.

5. Fluorine gas is bubbled into a solution of aluminum chloride.

6. Magnesium turnings are added to a solution of lead(II) acetate.

7. Iodine crystals are added to a solution of sodium chloride.

8. Calcium metal is added to a solution of nitrous acid.

9. A pea-sized piece of lithium is added to water.

10. A solution of iron(III) chloride is poured over a piece of platinum wire.

Note: On the AP reaction prediction section, all reactions "work"; in other words there will be no "No reactions" on the AP Exam.

Some possible questions for single replacement reactions:

- Identify evidence of a chemical change.
- Identify the substance being oxidized (reducing agent) or reduced (oxidizing agent).
- Predict the results for a burning splint test.
- Determine the total number of electrons lost/gained.
- Write the electron configuration for the element or ion.

Example   Magnesium turnings are added to 3.0 M hydrochloric acid.

If the reaction is done in a test tube and a burning splint is inserted, what will be observed?

Solution $Mg(s) + 2HCl(aq) \longrightarrow H_2(g) + MgCl_2(aq)$

$(Mg + 2H^+ \longrightarrow H_2 + Mg^{2+}$ net ionic equation)

The hydrogen gas will make sort of a "whoop" sound.

Example   A solution of copper(II) sulfate is spilled onto a sheet of freshly polished aluminum metal.

What is the total number of electrons transferred in this reaction?

Solution $2Al(s) + 3CuSO_4(aq) \longrightarrow 3Cu(s) + Al_2(SO_4)_3(aq)$

$(2Al + 3Cu^{2+} \longrightarrow 3Cu + 2Al^{3+}$ net ionic equation)

6 electrons are lost/gained

Example   A piece of zinc is placed in a solution of nickel(II) nitrate.

Write the complete electron configuration for the nickel(II) ion.

Solution $Zn(s) + Ni(NO_3)(aq) \longrightarrow Ni(s) + Zn(NO_3)_2(aq)$

$(Zn + Ni^{2+} \longrightarrow Ni + Zn^{2+}$ net ionic equation)

$1s^2 2s^2 2p^6 3s^2 3p^6 3d^8$ The $4s^2$ electrons are the valence electrons in a transition metal and are therefore removed before the $3d^8$ electrons in forming the Ni^{2+} ion.

Example   A bar of strontium metal is immersed in a 1.0 M copper(II) nitrate solution

What evidence of a chemical change would be observed in this reaction?

Solution $Sr(s) + Cu(NO_3)_2(aq) \longrightarrow Cu(s) + Sr(NO_3)_2(aq)$

$(Sr + Cu^{2+} \longrightarrow Cu + Sr^{2+}$ net ionic equation)

The blue copper(II) nitrate solution will become lighter and eventually become colorless. Also, reddish-brown copper will be deposited on the bar of strontium.

Notes

Chapter 9
Double Replacement (Metathesis) Reactions

In many reactions between two compounds in aqueous solution, the cations and anions appear to "switch partners" according to the following general equation:

$$AX + BY \longrightarrow AY + BX$$

The two compounds react to form two new compounds. No changes in oxidation numbers occur. Reactions of this type are known as *double replacement* or *metathesis reactions*. An example of such a reaction would be the mixing of aqueous solutions of potassium bromide and silver nitrate to form insoluble silver bromide (a precipitate) and aqueous potassium nitrate:

$$KBr(aq) + AgNO_3(aq) \longrightarrow AgBr(s) + KNO_3(aq)$$

Note that each cation pairs up with the anion in the other compound, thus switching partners. Anions do not pair up with anions and cations do not pair up with cations. Likes repel; opposites attract!

All double replacement reactions must have a "driving force" or a reason why the reaction will occur or "go to completion." The "driving force" in metathesis reactions is the removal of at least one pair of ions from solution.

This removal of ions can occur in one of three ways:

1. **Formation of a precipitate:** A precipitate is an insoluble substance (solid) formed by the reaction of two aqueous substances. It is the result of ions bonding together so strongly that the solvent (water) cannot pull them apart. The insoluble solid (or solids if a double precipitate occurs) will settle out (precipitate) from the solution and this results in the removal of ions from the solution.

2. **Formation of a gas:** Gases may form directly in a double replacement reaction or from the decomposition of one of the products. The gases will bubble off or evolve from the solution.

3. **Formation of primarily molecular species:** The formation of primarily unionized molecules in solution removes ions from the solution and the reaction "works" or is said to go to completion. Unionized or partially ionized molecules give solutions that are known as *nonelectrolytes* or *weak electrolytes*. The best known nonelectrolyte is water, which is formed in many acid–base neutralization reactions. Acetic acid is an example of an acid that is primarily molecular (weak electrolyte) when placed in water.

Reversible Reactions

If a double replacement reaction does not go to completion (no precipitate, gas or molecular species is formed), then the reaction is reversible (no ions have been removed). Reversible reactions are at equilibrium and have both forward and reverse reactions taking place. In a reversible reaction, evaporation of the water solvent will result in solid residues of both reactants and products. The reaction is not driven to completion (products) because no ions have been removed. A double arrow is used to designate a reversible reaction at equilibrium.

$$BaCl_2(aq) + 2NaNO_3(aq) \rightleftharpoons Ba(NO_3)_2(aq) + 2NaCl(aq)$$

Solubility Rules Table

Classifying ionic substances as soluble or insoluble based on their solubility in water is difficult. Nothing is completely "insoluble" in water. The degree of solubility also varies from one "soluble" substance to another. Nevertheless, the following solubility classification scheme is useful in predicting whether a possible double replacement reaction will go to completion. It is important to remember that any solubility "rules" should be regarded as an approximate guideline.

MAINLY WATER SOLUBLE

NO_3^-	All nitrates are soluble.
CH_3COO^- or $C_2H_3O_2^-$	All acetates are soluble except $AgCH_3COO^*$.
ClO_3^-	All chlorates are soluble.
Cl^-	All chlorides are soluble except $AgCl$, Hg_2Cl_2, $PbCl_2^*$.
Br^-	All bromides are soluble except $AgBr$, Hg_2Br_2 and $HgBr_2^*$, $PbBr_2^*$.
I^-	All iodides are soluble except AgI, Hg_2I_2, HgI_2, and PbI_2.
SO_4^{2-}	All sulfates are soluble except $Ag_2SO_4^*$, Hg_2SO_4, $PbSO_4$, $CaSO_4$, $SrSO_4^*$, and $BaSO_4$.
Alkali metal cations (Group 1) and NH_4^+	All are soluble.
H^+	All common inorganic acids and low molecular mass organic acids are soluble.

MAINLY WATER INSOLUBLE

CO_3^{2-}	All carbonates are insoluble except those of the IA elements and NH_4^+.
CrO_4^{2-}	All chromates are insoluble except those of the IA elements and NH_4^+ ions, as well as $CaCrO_4^*$ and $SrCrO_4^*$.
OH^-	All hydroxides are insoluble except those of the IA elements, NH_4^+, $Ba(OH)_2$, $Sr(OH)_2^*$, and $Ca(OH)_2^*$.
PO_4^{3-}	All phosphates are insoluble except those of the IA elements and NH_4^+.
SO_3^{2-}	All sulfites are insoluble except those of the IA elements and NH_4^+.
S^{2-}	All sulfides are insoluble except those of the IA and IIA elements and NH_4^+.

*Soluble compounds dissolve to the extent of at least 10 g/L at 25 °C. *Slightly soluble* compounds (marked with an *) dissolve in the range of from 1 g/L to 10 g/L at 25 °C. Those compounds that have a solubility of less than 1 g/L are considered to be *insoluble*. These standards are common but arbitrary.

Formation of a Precipitate

In order to predict double replacement reactions yielding precipitates, one **must memorize** the solubility rules listed on page 50.

Exercise 9–1: Predict and balance the following double replacement or metathesis reactions based on the solubility of the products. Use the abbreviations (aq) and (s) for the reactants and products. **All reactants are aqueous.** *Note:* Some of these reactions do not go to completion.

1. silver nitrate + potassium chromate

2. ammonium chloride + cobalt(II) sulfate

3. iron(III) sulfate + barium iodide

4. zinc acetate + cesium hydroxide

5. ammonium sulfide + lead(II) nitrate

6. lithium hydroxide + sodium chromate

7. chromium(III) bromide + sodium nitrate

8. rubidium phosphate + titanium(IV) nitrate

9. ammonium carbonate + nickel(II) chloride

10. tin(IV) nitrate + potassium sulfite

Note: Correct molecular formulas must be written for both the reactants and products before an equation may be balanced.

Formation of a Gas

Common gases formed in metathesis reactions are listed in the table below.

	Common Gases
H_2S	Any sulfide (salt of S^{2-}) plus any acid form $H_2S(g)$ and a salt.
CO_2	Any carbonate (salt of CO_3^{2-}) plus any acid form $CO_2(g)$, HOH, and a salt.
SO_2	Any sulfite (salt of SO_3^{2-}) plus any acid form $SO_2(g)$, HOH, and a salt.
NH_3	Any ammonium salt (salt of NH_4^+) plus any soluble strong hydroxide react upon heating to form $NH_3(g)$, HOH, and a salt.

Reactions that produce three of the gases (CO_2, SO_2, and NH_3) involve the initial formation of a substance that breaks down to give the gas and HOH.

Example 1   The reaction of Na_2SO_3 and HCl produces H_2SO_3:

$$Na_2SO_3(aq) + 2HCl(aq) \longrightarrow H_2SO_3(aq) + 2NaCl(aq)$$

Bubbling is observed in this reaction because the H_2SO_3 (sulfurous acid) is unstable and immediately decomposes to give HOH and SO_2 gas:

$$H_2SO_3(aq) \longrightarrow HOH(l) + SO_2(g)$$

The molecular equation for the overall or complete reaction, therefore, is:

$$Na_2SO_3(aq) + 2HCl(aq) \longrightarrow HOH(l) + SO_2(g) + 2NaCl(aq)$$

Example 2   A typical reaction of a carbonate and an acid is:

$$K_2CO_3(aq) + 2HNO_3(aq) \longrightarrow HOH(l) + CO_2(g) + 2KNO_3(aq)$$

Bubbling is also observed in this reaction. Theoretically H_2CO_3, carbonic acid, is formed, but the acid is unstable and immediately decomposes to form carbon dioxide gas and water according to the following equation:

$$H_2CO_3(aq) \longrightarrow HOH(l) + CO_2(g)$$

Example 3   Ammonium salts and soluble bases react as follows (particularly when the solution is warmed):

$$NH_4Cl(aq) + NaOH(aq) \longrightarrow NH_3(g) + HOH(l) + NaCl(aq)$$

The odor of ammonia gas is noted and moist red litmus paper held near the mouth of the container will turn blue. Theoretically NH_4OH, ammonium hydroxide, is produced (also known as ammonia water). The compound is unstable and decomposes into ammonia gas and water:

$$NH_4OH(aq) \longrightarrow NH_3(g) + HOH(l)$$

Example 4   The odor of rotten eggs and bubbling are noted when an acid is added to a sulfide. A typical reaction producing hydrogen sulfide gas is:

$$FeS(s) + 2HCl(aq) \longrightarrow FeCl_2(aq) + H_2S(g)$$

Helpful Tip: *Be aware of reactions involving the formation of carbon dioxide, sulfur dioxide, ammonia, and hydrogen sulfide gases on the AP Chemistry Examination. Over the years these reactions have appeared many, many times. Know these four gases and how they are produced!*

Exercise 9–2: Predict and balance the following double replacement or metathesis reactions. Use the abbreviations (s), (l), (g), and (aq) for the reactants and products. All reactants are aqueous unless otherwise stated. *Note:* Some of these reactions do not go to completion.

1. Ammonium sulfate and potassium hydroxide are mixed together.

2. Ammonium sulfide is reacted with hydrochloric acid.

3. Cobalt(II) chloride is combined with silver nitrate.

4. Solid calcium carbonate is reacted with sulfuric acid.

5. Potassium sulfite is reacted with hydrobromic acid.

6. Potassium sulfide is reacted with nitric acid.

7. ammonium iodide + magnesium sulfate

8. solid titanium(IV) carbonate + hydrochloric acid

9. solid calcium sulfite + acetic acid

10. strontium hydroxide + ammonium sulfide

Formation of a Molecular Species (Weak or Nonelectrolytes)

Formation of molecular products is the driving force for many metathesis reactions. Molecular products may include either partially dissociated (ionized) molecules, that is, weak electrolytes, or molecules that do not ionize or dissociate at all, that is, nonelectrolytes. Forming molecular products in double replacement reactions results in the removal of ions from solution. These types of reactions tend to go to completion (shift to the right) and primarily form products.

General list of rules:

A. The common strong acids $HClO_4$, $HClO_3$, HCl, HBr, HI, HNO_3, and H_2SO_4 are all strong electrolytes. **Memorize these seven strong acids!** All other common acids are weak acids and thus weak electrolytes. (CH_3COOH, H_3PO_4, HF, and HNO_2 are examples of weak acids.) *Note:* All organic acids (R–COOH) are weak electrolytes. All strong acids in their pure form (as opposed to dilute aqueous form) are nonelectrolytes (molecular). When water is added, the action of the solvent water with a strong acid produces a hydrated proton (hydronium ion) and a negatively charged anion. The process of making ions from molecular species is known as ionization. Strong acids ionize 100% in water. An example of a strong electrolyte undergoing ionization is as follows:

$$HCl(l) + H_2O(l) \longrightarrow H_3O^+(aq) + Cl^-(aq)$$

The overall reaction may be abbreviated as:

$$HCl(aq) \longrightarrow H^+(aq) + Cl^-(aq)$$

B. The common strong bases are the soluble hydroxides (those of Group 1 elements and Ba^{2+}) and the slightly soluble hydroxides (those of Ca^{2+} and Sr^{2+}). Strong bases, like strong acids, are strong electrolytes. **Memorize the strong bases!** NH_4OH is a soluble weak electrolyte that normally decomposes into $NH_3(g)$ and $HOH(l)$. Technically speaking, the pure compound ammonium hydroxide has never been isolated and the substance is more correctly known as aqueous ammonia. Most other hydroxides are insoluble. Pure *liquid* hydroxides are strong electrolytes because they already contain ions. The action of the solvent water releasing the ions of a base into solution is known as dissociation. Acids *ionize* in water; bases *dissociate!*

C. Most common (soluble) salts are strong electrolytes and thus dissociate into ions when placed in water.

D. Water is a weak electrolyte which is typically produced in acid–base neutralization reactions.

Some Examples of Weak Electrolytes Forming as Products (shown in bold)

$$Ca(CH_3COO)_2(aq) + 2HCl(aq) \longrightarrow CaCl_2(aq) + 2\mathbf{CH_3COOH(aq)}$$

$$2Na_3PO_4(aq) + 3H_2SO_4(aq) \longrightarrow 3Na_2SO_4(aq) + 2\mathbf{H_3PO_4(aq)}$$

$$HCl(aq) + NaOH(aq) \longrightarrow NaCl(aq) + \mathbf{HOH(l)}$$

Acid–Base Neutralization Reactions

Acids react with bases to produce salts and water. One mole of hydrogen ions will react with one mole of hydroxide ions to produce one mole of water. Learn which acids are strong acids (written in ionic form) and which are weak acids (written in molecular form). Check the solubility rules for the solubility of the salt produced. If it is soluble, it is written in ionic form; if it is insoluble it is written in molecular form. This will be covered further in Chapter 10.

$$\text{Acid + Base} \longrightarrow \text{Salt + Water}$$

The salt produced in the acid–base neutralization reaction consists of a cation from the base and an anion from the acid. An example is the salt sodium sulfate produced by the combination of sodium ions from sodium hydroxide and sulfate ions from sulfuric acid.

Example 1   Hydrogen sulfide gas is bubbled through excess potassium hydroxide solution.

$$H_2S(g) + 2KOH(aq) \longrightarrow K_2S(aq) + 2HOH(l)$$

Polyprotic acids can be tricky when it comes to predicting neutralization reactions. Sulfuric acid and phosphoric acid are classic examples frequently encountered on AP Examinations. If the base is in excess, all hydrogen ions will react with strong base to produce water.

Example 2   Dilute sulfuric acid is reacted with excess sodium hydroxide.

$$H_2SO_4(aq) + 2NaOH(aq) \longrightarrow Na_2SO_4(aq) + 2HOH(l)$$

If, however, this same reaction were described in terms of mixing equal numbers of moles of sulfuric acid and sodium hydroxide, then the coefficients for both reactants would be one and the salt that forms is sodium hydrogen sulfate (see Example 3).

Example 3   Equal number of moles of sulfuric acid and sodium hydroxide react.

$$H_2SO_4(aq) + NaOH(aq) \longrightarrow NaHSO_4(aq) + HOH(l)$$

As the following example demonstrates, it is important to take into account the quantity (concentration and amount) of each reactant.

Example 4   Equal volumes of 0.1 M phosphoric acid and 0.2 M sodium hydroxide are reacted together.

$$H_3PO_4(aq) + 2NaOH(l) \longrightarrow Na_2HPO_4(aq) + 2HOH(l)$$

Examples 5 and 6 show that in some reactions substances react with water before reacting further with an acid or a base. The acidic and basic anhydrides covered in Chapter 7 behave in such a manner. Watch for these substances that undergo two-step reactions.

Example 5   Excess sulfur dioxide gas is bubbled into a saturated solution of calcium hydroxide.

$$SO_2(g) + Ca(OH)_2(aq) \longrightarrow CaSO_3(s) + HOH(l)$$

SO_2 is an acid anhydride. If an acid + base yields a salt + water, then an **acid anhydride + basic anhydride will yield a salt.** *Note:* $SO_2(g)$ is the acid anhydride for sulfurous acid and $CaO(s)$ is the basic anhydride for calcium hydroxide.

Example 6   Sulfur dioxide gas and solid calcium oxide are reacted together.

$$SO_2(g) + CaO(s) \longrightarrow CaSO_3(s)$$

Exercise 9–3: Predict and balance the following reactions. Use the abbreviations (s), (l), (g), and (aq) for the reactants and products. All reactants are aqueous unless otherwise stated.

1. Carbon dioxide gas is bubbled through a solution of lithium hydroxide.

2. Sodium nitrite is reacted with hydrochloric acid.

3. ammonium bromide + sodium hydroxide

4. Carbon dioxide gas is reacted with solid potassium oxide.

5. Solid magnesium oxide is reacted with hydrochloric acid.

6. Equal numbers of moles of potassium hydroxide and phosphoric acid react.

7. Sodium fluoride reacts with dilute nitric acid.

8. ammonium carbonate + potassium bromide

9. Oxalic acid (0.1 M) reacts with an equal volume of 0.1 M cesium hydroxide.

10. silver nitrate + sodium chromate

Some possible questions for metathesis (double replacement) reactions:

- Identify evidence of a chemical change.
- Identify changes in color between reactants and products.
- Identify any spectator ions.
- Name the product(s).
- Describe any change in pH.

Example   Solid calcium sulfite is added to hydrochloric acid.

What evidence of a chemical reaction will be observed?

Solution $CaSO_3(s) + 2HCl(aq) \longrightarrow SO_2(g) + CaCl_2(aq) + H_2O(l)$

$(CaSO_3 + 2H^+ \longrightarrow SO_2 + Ca^{2+} + HOH$ net ionic equation$)$

Bubbles will be seen as sulfur dioxide gas is produced.

Example   Dilute hydrochloric acid is added to a solution of potassium carbonate.

If the products of this reaction are added to limewater (calcium hydroxide), what will be observed?
Explain.

Solution $K_2CO_3(aq) + 2HCl(aq) \longrightarrow CO_2(g) + H_2O(l) + 2KCl(aq)$

$(CO_3^{2-} + 2H^+ \longrightarrow CO_2 + HOH$ net ionic equation$)$

The limewater will turn a cloudy white as a precipitate of calcium carbonate forms.

Example   Solutions of silver nitrate and sodium chromate are mixed.

Name the precipitate obtained in this reaction.

Solution $Na_2CrO_4(aq) + 2AgNO_3(aq) \longrightarrow Ag_2CrO_4(s) + 2NaNO_3(aq)$

$(CrO_4^{2-} + 2Ag^+ \longrightarrow Ag_2CrO_4$ net ionic equation$)$

The precipitate is silver chromate.

Example   Solutions of silver nitrate and sodium chloride are mixed.

What is the color of the product?

Solution $AgNO_3(aq) + NaCl(aq) \longrightarrow AgCl(s) + NaNO_3(aq)$

$(Cl^- + Ag^+ \longrightarrow AgCl$ net ionic equation$)$

The silver chloride precipitate is white.

Example   Concentrated hydrochloric acid is added to solid manganese(II) sulfide.

What evidence of a chemical reaction will be detected?

Solution $2HCl(aq) + MnS(s) \longrightarrow H_2S(g) + MnCl_2(aq)$

                          $(2H^+ + MnS \longrightarrow H_2S + Mn^{2+}$ net ionic equation)

                          An odor of rotten eggs should be detected as the hydrogen sulfide gas is emitted.

Example   Equal volumes of equimolar phosphoric acid and potassium hydroxide solutions
                          are mixed.

Will the pH of the phosphoric acid solution increase, decrease or remain the same as the potassium
hydroxide is added to it?

Solution $H_3PO_4(aq) + KOH(aq) \longrightarrow KH_2PO_4(aq) + HOH(l)$

                          $(H_3PO_4 + OH^- \longrightarrow H_2PO_4^- + HOH$ net ionic equation)

                          The pH of the phosphoric acid solution will increase (it will become less acidic or more
                          basic) as the acid is neutralized by the base (KOH).

Example   Equal volumes of 0.10 M hydrobromic acid and 0.10 M sodium hydroxide are
                          combined.

If this reaction was carried out and then a drop of bromthymol blue was added to the final solution, what
color would be observed? Explain.

Solution $NaOH(aq) + HBr(aq) \longrightarrow HOH(l) + NaBr(aq)$

                          $(OH^- + H^+ \longrightarrow HOH$ net ionic equation)

                          The final color of the solution would be green, which is the neutral color of the
                          indicator. The neutralization reaction produces water (which is neutral) and a neutral
                          salt (NaBr). Bromthymol blue is yellow in acidic solutions and blue in basic solutions.

Example   Ammonium chloride crystals are added to a solution of sodium hydroxide in a test tube.

If a piece of moist litmus paper is held at the mouth of the test tube, what would be observed? Explain.

Solution $NaOH(aq) + NH_4Cl(s) \longrightarrow NH_3(g) + HOH(l) + NaCl(aq)$

                          $(OH^- + NH_4Cl \longrightarrow NH_3 + HOH + Cl^-$ net ionic equation)

                          The NH_3 vapor will turn the litmus paper blue because it is a weak base.

Chapter 10
Aqueous Solutions and Ionic Equations

Finally, after working through the first nine chapters, it is time to learn how to write chemical equations in the form required on the AP Chemistry Examination! All equations in previous chapters of this book were written as if the reactants and products were molecular. No explicit indication was made that soluble compounds had actually dissociated or ionized into ions. Molecular equations were used to represent all reactants and products by means of their complete formulas—how a reactant or product actually exists in solution was ignored up to this point. For example, given the reaction between aqueous solutions of cadmium nitrate and sodium sulfide, the following molecular equation results:

$$Cd(NO_3)_2(aq) + Na_2S(aq) \longrightarrow CdS(s) + 2NaNO_3(aq)$$

Officially, on the AP Chemistry Examination, the phase symbols (aq), (s), (l), and (g) are not required to be written in an equation. These symbols are frequently used as an aid for beginning chemistry students. Now we are ready to move on to a more advanced level of equation writing! Writing equations for the AP Chemistry Exam requires understanding overall and net ionic equations.

Overall (Total) Ionic Equations

In overall or total ionic equations, formulas of the reactants and products are written to show the predominant form of each substance as it exists in aqueous solution. Soluble salts, as well as strong acids and strong bases are written as separated ions in overall ionic equations. Insoluble salts, suspensions, solids, weak acids, weak bases, gases, water, and organic compounds are always written as individual molecules in overall ionic equations. The overall ionic equation for the cadmium nitrate/sodium sulfide reaction given above would be:

$$Cd^{2+}(aq) + 2NO_3^-(aq) + 2Na^+(aq) + S^{2-}(aq) \longrightarrow CdS(s) + 2Na^+(aq) + 2NO_3^-(aq)$$

Note that this equation illustrates the following facts: (1) cadmium nitrate dissociates into three ions (one cadmium ion and two nitrate ions); (2) sodium sulfide dissociates into two sodium ions and one sulfide ion; (3) the soluble sodium nitrate formed remains dissociated as two sodium and two nitrate ions; and (4) the precipitated (insoluble) cadmium sulfide is undissociated. Also note that the parentheses that appear in a molecular formula are not used when representing the ionic form in solution, e.g., dissociated cadmium nitrate contains no parentheses.

The following questions should help to determine if formula units should be written as separate ions or as uncharged "molecules" in an overall ionic equation:

* Does the substance dissolve in water? If no, write the molecular formula. If yes, see below.
* If the substance dissolves in water, does it dissociate into either a salt or a strong base? If yes, write the ionic form of the formula; if no, write the molecular formula.
* If the substance dissolves in water, does it ionize into a strong acid? If yes, then write the ionic form of the formula; if no, write the molecular formula.

The only common substances that should be written as ions in ionic equations are:

| • Soluble Salts | • Strong Acids | • Strong Bases |

It should now be obvious to the reader why tables of solubility rules, strong acids, and strong bases must be memorized, along with the formulas for polyatomic ions. The rules that were presented in earlier chapters **must** be memorized. There is no other way to learn the necessary rules. Practice, practice, and more practice is necessary in order to become good at writing ionic equations. Also, remember that the phase symbols (aq), (s), (l), and (g) are not officially required to be included in an equation. During the AP Examination, it is a waste of time to include these phase symbols. Get in the habit of not writing phase symbols!

The following examples should provide a better feel for writing overall ionic equations. If writing an overall ionic equation is causing you trouble, try writing the molecular equation first.

Example 1. Aqueous solutions of sulfuric acid and excess sodium hydroxide are reacted.

Molecular Eqn. $\quad H_2SO_4(aq) + 2NaOH(aq) \longrightarrow 2HOH(l) + Na_2SO_4(aq)$

Overall Ionic Eqn. $\quad H^+ + HSO_4^- + 2Na^+ + 2OH^- \longrightarrow 2HOH + 2Na^+ + SO_4^{2-}$

> **Special Note:** The first proton in sulfuric acid is ionized completely; the second proton is only partially ionized. Sulfuric acid is the only polyprotic acid that exhibits this property. All other polyprotic acids are weak acids and are therefore written in their molecular forms. For sulfuric acid, get in the habit of writing: $H_2SO_4 \longrightarrow H^+ + HSO_4^-$

Example 2. Aqueous solutions of sulfuric acid and excess barium hydroxide are combined.

Molecular Eqn. $\quad H_2SO_4(aq) + Ba(OH)_2(aq) \longrightarrow 2HOH(l) + BaSO_4(s)$

Overall Ionic Eqn. $\quad H^+ + HSO_4^- + Ba^{2+} + 2OH^- \longrightarrow 2HOH + BaSO_4$

Example 3. Aqueous hydrochloric acid reacts with solid iron(II) sulfide.

Molecular Eqn. $\quad 2HCl(aq) + FeS(s) \longrightarrow H_2S(g) + FeCl_2(aq)$

Overall Ionic Eqn. $\quad 2H^+ + 2Cl^- + FeS \longrightarrow H_2S + Fe^{2+} + 2Cl^-$

Example 4. Excess aqueous acetic acid reacts with aqueous sodium sulfite.

Molecular Eqn. $\quad 2CH_3COOH(aq) + Na_2SO_3(aq) \longrightarrow 2NaCH_3COO(aq) + HOH(l) + SO_2(g)$

Overall Ionic Eqn. $\quad 2CH_3COOH + 2Na^+ + SO_3^{2-} \longrightarrow 2Na^+ + 2CH_3COO^- + HOH + SO_2$

Example 5. Calcium carbonate in aqueous suspension reacts with dilute hydrochloric acid.

Molecular Eqn. $\quad CaCO_3(s) + 2HCl(aq) \longrightarrow CaCl_2(aq) + HOH(l) + CO_2(g)$

Overall Ionic Eqn. $\quad CaCO_3 + 2H^+ + 2Cl^- \longrightarrow Ca^{2+} + 2Cl^- + HOH + CO_2$

Example 6. Aqueous ammonium sulfide reacts with excess aqueous lithium hydroxide.

Molecular Eqn. $\quad (NH_4)_2S(aq) + 2LiOH(aq) \longrightarrow Li_2S(aq) + 2NH_3(g) + 2HOH(l)$

Overall Ionic Eqn. $\quad 2NH_4^+ + S^{2-} + 2Li^+ + 2OH^- \longrightarrow 2Li^+ + S^{2-} + 2NH_3 + 2HOH$

Example 7. Aqueous solutions of calcium nitrate and rubidium chloride are mixed together.

Molecular Eqn. $\quad Ca(NO_3)_2(aq) + 2RbCl(aq) \rightleftharpoons CaCl_2(aq) + 2RbNO_3(aq)$

Overall Ionic Eqn. $\quad Ca^{2+} + 2NO_3^- + 2Rb^+ + 2Cl^- \rightleftharpoons Ca^{2+} + 2Cl^- + 2Rb^+ + 2NO_3^-$

Example 8. Aqueous solutions of potassium chromate and silver nitrate are reacted together.

Molecular Eqn. $\quad K_2CrO_4(aq) + 2AgNO_3(aq) \longrightarrow 2KNO_3(aq) + Ag_2CrO_4(s)$

Overall Ionic Eqn. $\quad 2K^+ + CrO_4^{2-} + 2Ag^+ + 2NO_3^- \longrightarrow 2K^+ + 2NO_3^- + Ag_2CrO_4$

Example 9. Equal volumes of 0.2 M potassium hydroxide and 0.2 M phosphoric acid are reacted.

Molecular Eqn. $\quad\quad\quad$ $KOH(aq) + H_3PO_4(aq) \longrightarrow HOH(l) + KH_2PO_4(aq)$

Overall Ionic Eqn. $\quad\;$ $K^+ + OH^- + H_3PO_4 \longrightarrow HOH + K^+ + H_2PO_4^-$

Example 10. Methane gas is burned completely in air.

Molecular Eqn. $\quad\quad\quad$ $CH_4(g) + 2O_2(g) \longrightarrow 2HOH(g) + CO_2(g)$

Overall Ionic Eqn. $\quad\;$ The molecular equation is also the ionic equation for this reaction. No ions are present; all reactants and products are molecular.

Example 11. Sodium acetate undergoes hydrolysis when placed in water.

Molecular Eqn. $\quad\quad\quad$ $NaCH_3COO(aq) + HOH(l) \longrightarrow NaOH(aq) + CH_3COOH(aq)$

Overall Ionic Eqn. $\quad\;$ $Na^+ + CH_3COO^- + HOH \longrightarrow Na^+ + OH^- + CH_3COOH$

Example 12. Solid calcium chlorate is heated.

Molecular Eqn. $\quad\quad\quad$ $Ca(ClO_3)_2(s) \longrightarrow CaCl_2(s) + 3O_2(g)$

Overall Ionic Eqn. $\quad\;$ The molecular equation is also the ionic equation for this reaction because no ions are present.

Hydrolysis Reactions

Example 11 in the previous section dealt with the hydrolysis of sodium acetate when it is placed in water. The reaction of a salt with water to form molecular species is known as *hydrolysis*. Previously, a neutralization reaction was described as an acid plus a base yields a salt plus water. Hydrolysis is the reverse of neutralization and results when a salt plus water yields an acid plus a base. Neutralization and hydrolysis reactions are both examples of metathesis reactions. Hydrolysis problems, however, deal with the formation of a weak acid and/or a weak base. Salts are a product of neutralization, but salts that undergo hydrolysis usually are not neutral. Be aware of this!

General Hydrolysis Reaction $\quad\quad$ salt + HOH $\longrightarrow$ acid + base

$$MA + HOH \longrightarrow HA + MOH$$

Example 1. Aqueous ammonium chloride undergoes hydrolysis when placed in water.

$$NH_4^+ + Cl^- + HOH \longrightarrow H^+ + Cl^- + NH_4OH$$

Ammonium chloride is a salt made from a strong acid (hydrochloric) and a weak base (ammonium hydroxide). In an aqueous solution, ammonium chloride will hydrolyze to give molecular ammonium hydroxide (a weak base) and ionic hydrochloric acid (a strong acid). Ammonium chloride hydrolyzes to give an acid solution because the free hydronium ions in solution outnumber the free hydroxide ions in solution ($H^+ > OH^-$). Molecular ammonium hydroxide has tied up most of the hydroxide ions.

Key Point: Salts of a strong acid + a weak base will always hydrolyze to give an acidic solution.

Example 2. Aqueous potassium fluoride undergoes hydrolysis when placed in water.

$$K^+ + F^- + HOH \longrightarrow K^+ + OH^- + HF$$

Potassium fluoride is a salt made from a weak acid (hydrofluoric) and a strong base (potassium hydroxide). In an aqueous solution, potassium fluoride will hydrolyze to give molecular hydrofluoric acid (a weak acid) and ionic potassium hydroxide (a strong base). Potassium fluoride hydrolyzes to give a basic solution because the free hydroxide ions in solution outnumber the available free hydronium ions in solution ($H^+ < OH^-$). Molecular hydrofluoric acid has tied up most of the hydronium ions.

> Key Point: Salts of a weak acid + a strong base will always hydrolyze to give a basic solution.

Example 3. Sodium chloride and water are mixed together.

$$Na^+ + Cl^- + HOH \longrightarrow Na^+ + OH^- + H^+ + Cl^-$$

Sodium chloride is a salt produced from the neutralization of a strong acid (hydrochloric) and a strong base (sodium hydroxide). In an aqueous solution, sodium chloride does not undergo hydrolysis; no molecular acid or base is produced. The net result is a neutral solution, where $H^+ = OH^-$. Neither the hydronium nor hydroxyl ions are tied up by a weak electrolyte.

> Key Point: Salts of a strong acid + a strong base do not undergo
> hydrolysis. Their aqueous solutions are always neutral.

Example 4. Ammonium fluoride and water are mixed together.

$$NH_4^+ + F^- + HOH \longrightarrow HF + NH_4OH$$

Ammonium fluoride is a salt produced by the neutralization of a weak acid (hydrofluoric) and a weak base (ammonium hydroxide). In an aqueous solution, ammonium fluoride will hydrolyze to give molecular hydrofluoric acid and molecular ammonium hydroxide. Reacting the salt of a weak acid and a weak base results in a solution that could be acidic, basic or neutral. The final hydronium ion concentration depends on which is stronger, the weak acid or the weak base. One has to appeal to K_a and K_b values found in *Acid/ Base Ionization Constant* tables. In the case of ammonium fluoride, the hydrofluoric acid is a weak acid, but the ammonium hydroxide is a slightly weaker base. Therefore, in this reaction the ammonium hydroxide will hydrolyze more (form more molecules) than hydrofluoric acid. The solution that results will be slightly acidic ($H^+ > OH^-$).

> Key Point: Salts of a weak acid + a weak base may hydrolyze to give an acidic, basic
> or neutral solution. The final result depends on the K_a and K_b values of the acid
> and base formed during the hydrolysis process.

- If the weak acid formed in hydrolysis is stronger than the weak base formed, the overall solution will be slightly acidic.

- If the weak base formed in hydrolysis is stronger than the weak acid formed, the overall solution will be slightly alkaline (basic).

- If the weak acid and base formed during hydrolysis are of equal strength, the overall solution will be neutral.

Hopefully, the advantage of writing overall ionic equations is now apparent. In the case of hydrolysis reactions, the overall ionic equation allows us to "see" what is happening in the solution. A word of caution, however, is necessary because hydrolysis reactions produce very small amounts of acid and base in solution. For the most part, a salt plus water will remain primarily as a salt plus water. Hydrolysis is a limited reaction that forms small quantities of hydronium and hydroxyl ions in solution. If ammonium hydroxide, carbonic acid, or sulfurous acid are formed in a hydrolysis reaction, the concentrations that result are so small that their decomposition into water and the corresponding gas does not take place.

Exercise 10–1: Write balanced overall ionic equations for the following reactions.

1. aqueous nickel(II) nitrate + aqueous cesium hydroxide

2. equal volumes of equal molar concentrations of sulfuric acid and sodium hydroxide

3. Solid potassium chlorate is strongly heated.

4. potassium tartrate solution + water

5. Solid lithium metal is added to water.

6. Aqueous solutions of magnesium nitrate and sodium bromide are mixed together.

7. aqueous solutions of oxalic acid and excess potassium hydroxide

8. solid cobalt(II) hydroxide + hydroiodic acid

9. aqueous solution of manganese(II) sulfate undergoing hydrolysis

10. aqueous sodium carbonate + chlorous acid

11. aqueous solutions of potassium phosphate and excess hydrobromic acid

Exercise 10–2: Write balanced overall ionic equations for all reactions in **Exercise 7–2.** If no overall *ionic* equation exists for a reaction, write *none*.

Exercise 10–3: Write balanced overall ionic equations for all reactions in **Exercise 8–1.** If no overall *ionic* equation exists for a reaction, write *none*.

Exercise 10–4: Write balanced overall ionic equations for all reactions in **Exercise 9–1.** If no overall *ionic* equation exists for a reaction, write *none*.

Exercise 10–5: Write balanced overall ionic equations for all reactions in **Exercise 9–2.** If no overall *ionic* equation exists for a reaction, write *none*.

Exercise 10–6: Write balanced overall ionic equations for all reactions in **Exercise 9–3.** If no overall *ionic* equation exists for a reaction, write *none*.

Net Ionic Equations

The topic of ionic equations is not quite finished—*net ionic equations* need to be discussed. A *net ionic equation* is written to show only the species that react or undergo change in aqueous solution. The *net ionic equation* is obtained by eliminating the *spectator ions* from the overall ionic equation. All that is left are the ions that have changed chemically. Spectators at a sporting event watch the action unfolding in front of them rather than participating; spectator ions likewise do not participate in the reaction. The elimination of spectator ions allows us to concentrate only on the reacting species.

Molecular equations, as discussed in Chapter 7, provide complete chemical formulas that are needed to carry out stoichiometry calculations. Overall ionic equations, the intermediate between molecular and net ionic equations, show what is happening to all species in the solution. Such equations are very helpful when dealing with hydrolysis, electrical conductivity, and colligative properties. Net ionic equations are the simplest form of equations and show only the reacting species.

Let's go back to the example at the very beginning of this chapter—the reaction between aqueous solutions of cadmium nitrate and sodium sulfide:

Word Eqn.   cadmium nitrate + sodium sulfide $\longrightarrow$ cadmium sulfide + sodium nitrate

Molecular Eqn. $Cd(NO_3)_2(aq) + Na_2S(aq) \longrightarrow CdS(s) + 2NaNO_3(aq)$

Overall Ionic Eqn. $Cd^{2+}(aq) + \mathbf{2NO_3^-(aq)} + \mathbf{2Na^+(aq)} + S^{2-}(aq) \longrightarrow CdS(s) + \mathbf{2Na^+(aq)} + \mathbf{2NO_3^-(aq)}$
                     (The spectator ions have been highlighted in bold.)

Net Ionic Eqn. $Cd^{2+} + S^{2-} \longrightarrow CdS$
                     (Remember, spectator ions do not undergo any change in a reaction and are eliminated from the net ionic equation.)

Examples of Net Ionic Equations

In order to gain practice with the proper writing of net ionic equations, several examples from Chapters 7, 8, and 10 are reproduced below showing both the overall and net ionic equations, as appropriate.

Chapter 7 Examples

Example 1. Solid sodium oxide is added to water.

   Overall Ionic Eqn. $Na_2O(s) + HOH(l) \longrightarrow 2Na^+(aq) + 2OH^-(aq)$
   Net Ionic Eqn. $Na_2O(s) + HOH(l) \longrightarrow 2Na^+(aq) + 2OH^-(aq)$

Example 2. Solid magnesium oxide is added to water.

   Overall Ionic Eqn.   The ionic equation for this reaction is the same as the molecular equation. No ions are present; $Mg(OH)_2$ is a weak base.
   Net Ionic Eqn.   No ions are present.

Example 3. Carbon dioxide is bubbled into water.

   Overall Ionic Eqn.   The ionic equation for this reaction is the same as the molecular equation. No ions are present; H_2CO_3 is a weak acid.
   Net Ionic Eqn.   No ions are present.

Example 4. Dinitrogen pentoxide is bubbled into water.

Overall Ionic Eqn. $N_2O_5(g) + HOH(l) \longrightarrow 2H^+(aq) + 2NO_3^-(aq)$
Net Ionic Eqn. $N_2O_5(g) + HOH(l) \longrightarrow 2H^+(aq) + 2NO_3^-(aq)$

Chapter 8 Examples

Example 5. Magnesium turnings are added to a solution of iron(III) chloride.

Overall Ionic Eqn. $3Mg(s) + Fe^{3+}(aq) + 6Cl^-(aq) \longrightarrow 3Mg^{2+}(aq) + 6Cl^-(aq) + 2Fe(s)$
Net Ionic Eqn. $3Mg(s) + 2Fe^{3+}(aq) \longrightarrow 3Mg^{2+}(aq) + 2Fe(s)$

Example 6. Sodium is added to water.

Overall Ionic Eqn. $2Na(s) + 2HOH(l) \longrightarrow H_2(g) + 2Na^+(aq) + 2OH^-(aq)$
Net Ionic Eqn. $2Na(s) + 2HOH(l) \longrightarrow H_2(g) + 2Na^+(aq) + 2OH^-(aq)$

Example 7. Lithium is added to hydrochloric acid.

Overall Ionic Eqn. $2Li(s) + 2H^+(aq) + 2Cl^-(aq) \longrightarrow H_2(g) + 2Li^+(aq) + 2Cl^-(aq)$
Net Ionic Eqn. $2Li(s) + 2H^+(aq) \longrightarrow H_2(g) + 2Li^+(aq)$

Example 8. Chlorine gas is bubbled into a solution of potassium iodide.

Overall Ionic Eqn. $Cl_2(g) + 2K^+(aq) + 2I^-(aq) \longrightarrow 2K^+(aq) + 2Cl^-(aq) + I_2(s)$
Net Ionic Eqn. $Cl_2(g) + 2I^-(aq) \longrightarrow 2Cl^-(aq) + I_2(s)$

Chapter 10 Examples

Example 1. Aqueous solutions of sulfuric acid and excess sodium hydroxide are reacted.

Overall Ionic Eqn. $H^+ + HSO_4^- + 2Na^+ + 2OH^- \longrightarrow 2HOH + 2Na^+ + SO_4^{2-}$
Net Ionic Eqn. $H^+ + HSO_4^- + 2OH^- \longrightarrow 2HOH + SO_4^{2-}$

Example 2. Aqueous solutions of sulfuric acid and excess barium hydroxide.

Overall Ionic Eqn. $H^+ + HSO_4^- + Ba^{2+} + 2OH^- \longrightarrow 2HOH + BaSO_4$
Net Ionic Eqn. $H^+ + HSO_4^- + Ba^{2+} + 2OH^- \longrightarrow 2HOH + BaSO_4$

Example 3. Aqueous hydrochloric acid reacts with solid iron(II) sulfide.

Overall Ionic Eqn. $2H^+ + 2Cl^- + FeS \longrightarrow H_2S + Fe^{2+} + 2Cl^-$
Net Ionic Eqn. $2H^+ + FeS \longrightarrow H_2S + Fe^{2+}$

Example 4. Excess aqueous acetic acid reacts with aqueous sodium sulfite.

Overall Ionic Eqn. $2CH_3COOH + 2Na^+ + SO_3^{2-} \longrightarrow 2Na^+ + 2CH_3COO^- + HOH + SO_2$
Net Ionic Eqn. $2CH_3COOH + SO_3^{2-} \longrightarrow 2CH_3COO^- + HOH + SO_2$

Example 5. Calcium carbonate in aqueous suspension reacts with dilute hydrochloric acid.

Overall Ionic Eqn. $CaCO_3 + 2H^+ + 2Cl^- \longrightarrow Ca^{2+} + 2Cl^- + HOH + CO_2$
Net Ionic Eqn. $CaCO_3 + 2H^+ \longrightarrow Ca^{2+} + HOH + CO_2$

Example 6. Aqueous ammonium sulfide reacts with excess aqueous lithium hydroxide.

Overall Ionic Eqn. $2NH_4^+ + S^{2-} + 2Li^+ + 2OH^- \longrightarrow 2Li^+ + S^{2-} + 2NH_3 + 2HOH$

Net Ionic Eqn. $2NH_4^+ + 2OH^- \longrightarrow 2NH_3 + 2HOH$ *or*

$NH_4^+ + OH^- \longrightarrow NH_3 + HOH$

Example 7. Aqueous solutions of calcium nitrate and rubidium chloride are mixed together.

Overall Ionic Eqn. $Ca^{2+} + 2NO_3^- + 2Rb^+ + 2Cl^- \rightleftharpoons Ca^{2+} + 2Cl^- + 2Rb^+ + 2NO_3^-$

Net Ionic Eqn.   None because all ions are spectators!

Example 8. Aqueous solutions of potassium chromate and silver nitrate react.

Overall Ionic Eqn. $2K^+ + CrO_4^{2-} + 2Ag^+ + 2NO_3^- \longrightarrow 2K^+ + 2NO_3^- + Ag_2CrO_4$

Net Ionic Eqn. $2Ag^+ + CrO_4^{2-} \longrightarrow Ag_2CrO_4$

Example 9. Equal volumes of 0.2 M potassium hydroxide and 0.2 M phosphoric acid are reacted together.

Overall Ionic Eqn. $K^+ + OH^- + H_3PO_4 \longrightarrow HOH + K^+ + H_2PO_4^-$

Net Ionic Eqn. $OH^- + H_3PO_4 \longrightarrow HOH + H_2PO_4^-$

Example 10. Methane gas is completely burned in air.

Overall Ionic Eqn.   The ionic equation for this reaction is the same as the molecular equation. No ions are present; all reactants and products are molecular! Only a molecular equation can be written!

Net Ionic Eqn.   No ions are present. Only a molecular equation can be written!

Example 11. Sodium acetate undergoes hydrolysis when placed in water.

Overall Ionic Eqn. $Na^+ + CH_3COO^- + HOH \longrightarrow Na^+ + OH^- + CH_3COOH$

Net Ionic Eqn. $CH_3COO^- + HOH \longrightarrow OH^- + CH_3COOH$

Example 12. Solid calcium chlorate is heated.

Overall Ionic Eqn.   Same as molecular equation.

Net Ionic Eqn.   Same as molecular equation.

Overall ionic and net ionic equations are used a great deal of the time in advanced chemistry work. Therefore it is important to learn how to skillfully write balanced overall ionic and net ionic equations. The AP Chemistry Examination has an equation section worth 15 points that is worth 10% of the free-response section. **All equations in this section of the AP Exam must be written as balanced net ionic equations!**

Exercise 10–7:   Write balanced net ionic equations for all reactions in **Exercise 10–1.** Write *none* if a *net ionic* equation does not exist.

Exercise 10–8:   Write balanced net ionic equations for all reactions in **Exercise 7–2.** Write *none* if a *net ionic* equation does not exist.

Exercise 10–9:   Write balanced net ionic equations for all reactions in **Exercise 8–1.** Write *none* if a *net ionic* equation does not exist.

Exercise 10–10:   Write balanced net ionic equations for all reactions in **Exercise 9–1.** Write *none* if a *net ionic* equation does not exist.

Exercise 10–11:   Write balanced net ionic equations for all reactions in **Exercise 9–2.** Write *none* if a *net ionic* equation does not exist.

Exercise 10–12:   Write balanced net ionic equations for all reactions in **Exercise 9–3.** Write *none* if a *net ionic* equation does not exist.

Notes

Chapter 11
Balancing Equations for Redox Reactions

Reactions involving the processes of oxidation and reduction are quite common. They can range from simple combinations of elements such as the formation of rust to more complex reactions that result in a silverplated piece of jewelry. Living things rely on the oxidation of organic compounds to provide them with energy either internally during digestion or externally during the burning of fuel.

Oxidation and reduction reactions go together—one cannot occur without the other. Oxidation occurs when electrons are lost by one substance and reduction involves the gain of electrons by another substance. Since these processes always occur together, chemists often refer to them as oxidation–reduction or "redox" reactions.

Balancing Redox Equations

In order to balance redox equations, the first step is to determine the substance that has been oxidized and the one reduced. This can be done by assigning oxidation numbers to each atom in the reaction. An increase in oxidation number (becoming more positive) indicates that the element is being *oxidized*. A decrease in oxidation number (becoming less positive or more negative) indicates that the element is being *reduced*. Review the rules for assigning oxidation numbers as outlined in Chapter 3.

There are two basic methods that are commonly used when balancing redox equations—the *oxidation states method* and the *ion–electron method*. Many simple redox equations can be balanced using the oxidation states method. Many redox reactions, especially ones that may be encountered in the multiple choice or reaction prediction section of the AP Exam, are a bit more complicated. Here, the ion–electron method should be used. A third method that may be introduced by your textbook or your teacher is the *half-reaction method*. Whatever method you choose, clues can often be found to identify a redox reaction by noticing key words such as *"acidic"* or *"alkaline"*; balancing these equations requires a few more steps. Notice that spectator ions have been omitted in the examples and practice exercises. Ultimately, a balanced redox equation will be balanced in terms of mass and charge.

Oxidation States Method

1. Assign oxidation numbers to each atom in the equation.
2. Identify the elements being oxidized and reduced, respectively, and use coefficients to balance those redox atoms.
3. Determine the number of electrons lost and gained by the redox atoms.
4. Use coefficients to balance the electrons lost and gained.
5. Use coefficients to balance the nonredox atoms by inspection.

Example 1. Cadmium sulfide is reacted with iodine and hydrogen chloride to produce cadmium chloride, hydrogen iodide, and sulfur.

$$CdS + I_2 + HCl \longrightarrow CdCl_2 + HI + S$$

1. *Oxidation #s* $\quad Cd^{2+} S^{2-} + I_2^0 + H^+ Cl^- \longrightarrow Cd^{2+} Cl_2^- + H^+ I^- + S^0$

2. *Redox atoms*	Sulfur changes from 2– to 0; sulfur is oxidized. Iodine changes from 0 to 1–; iodine is reduced. Because iodine is diatomic on the reactant side and both atoms must be reduced, there must be two HI molecules formed on the product side.

$$CdS + I_2 + HCl \longrightarrow CdCl_2 + 2HI + S$$

3. *e⁻s lost/gained*	Since sulfur loses 2 electrons and each iodine atom gains one (for a total of 2), the number of electrons lost and gained is equal.
4. *Balance e⁻s*	No additional coefficients are required to balance electrons.
5. *Balance atoms*	Finally, the hydrogen and chlorine atoms must be balanced.

$$CdS + I_2 + 2HCl \longrightarrow CdCl_2 + 2HI + S$$

Example 2. Chlorine gas is reacted with calcium hydroxide to produce calcium chloride, calcium chlorate, and water.

$$Cl_2 + Ca(OH)_2 \longrightarrow CaCl_2 + Ca(ClO_3)_2 + H_2O$$

1. *Oxidation #s*	$Cl_2{}^0 + Ca^{2+}(O^{2-}H^+)_2 \longrightarrow Ca^{2+}Cl_2{}^- + Ca^{2+}(Cl^{5+}O^{2-}{}_3)_2 + H^+{}_2O^{2-}$
2. *Redox atoms*	Chlorine changes from 0 to 5+; chlorine is oxidized. Chlorine changes from 0 to 1–; chlorine is reduced. Notice that chlorine is diatomic on the reactant side and both calcium chloride and calcium chlorate involve 2 chlorine atoms per formula unit. Because all chlorines involved are in pairs, no balancing needs to be done at this point.
3. *e⁻s lost/gained*	Since a total of 10 electrons are being lost (0 to 5+, a loss of 5 electrons times two) in the oxidation step, 10 electrons must also be gained in the reduction process.
4. *Balance e⁻s*	Therefore, the product obtained in the reduction process must be multiplied by 5.

$$5Cl_2 + Ca(OH)_2 \longrightarrow 5CaCl_2 + Ca(ClO_3)_2 + H_2O$$

5. *Balance atoms*	Finish balancing calcium, chlorine, oxygen, and hydrogen.

$$6 Cl_2 + 6Ca(OH)_2 \longrightarrow 5CaCl_2 + Ca(ClO_3)_2 + 6H_2O$$

In the final balanced equation for this reaction five chlorine molecules have been reduced and one chlorine molecule has been oxidized. This example also illustrates the concept of a *disproportionation reaction*, a reaction in which the same substance is being both oxidized and reduced. In the case above—chlorine!

Ion–Electron Method in Acidic Solution

1. Assign oxidation numbers to each atom in the equation.
2. Identify what is being oxidized and reduced.
3. Divide the skeleton equation into two half-reactions—one for the oxidation process and one for the reduction process.
4. For each half-reaction, balance atoms other than H and O.
5. For each half-reaction, balance oxygen by adding H_2O to the side that needs O and balance hydrogen by adding H^+ to the side that needs H.

6. For each half-reaction, add up the total charge on each side and add electrons on either side as needed to balance the total charge on both sides of each half-reaction. (You should notice that the number of electrons corresponds to the change in oxidation state for each half-reaction.)

7. Balance the number of electrons being lost and gained.

8. Add the two half-reactions and cancel anything that is the same on both sides.

Example. $MnO_4^- + SO_2 \longrightarrow Mn^{2+} + SO_4^{2-}$ *under acidic conditions*

1. *Oxidation #s* $(Mn^{7+}O^{2-}_4)^- + (S^{4+}O^{2-}_2) \longrightarrow Mn^{2+} + (S^{6+}O^{2-}_4)^{2-}$

2. *Redox*   Mn is being reduced and S is being oxidized.

3. *Half-reactions* $MnO_4^- \longrightarrow Mn^{2+}$ *reduction*
                              $SO_2 \longrightarrow SO_4^{2-}$ *oxidation*

4. *Balance atoms*   No atoms other than H and O require balancing.

5. *Balance H & O* $MnO_4^- + 8H^+ \longrightarrow Mn^{2+} + 4H_2O$
                              $SO_2 + 2H_2O \longrightarrow SO_4^{2-} + 4H^+$

6. *Net charge* $MnO_4^- + 8H^+ \longrightarrow Mn^{2+} + 4H_2O$
                         $\qquad\quad +7 \qquad\qquad +2$

   $MnO_4^- + 8H^+ + 5e^- \longrightarrow Mn^{2+} + 4H_2O$

   $SO_2 + 2H_2O \longrightarrow SO_4^{2-} + 4H^+$
   $\quad\; 0 \qquad\qquad\quad +2$

   $SO_2 + 2H_2O \longrightarrow SO_4^{2-} + 4H^+ + 2e^-$

7. *Balance electrons*   To balance electrons, the first half-reaction must be multiplied by 2 and the second half-reaction by 5.

   $2MnO_4^- + 16H^+ + 10e^- \longrightarrow 2Mn^{2+} + 8H_2O$
   $5SO_2 + 10H_2O \longrightarrow 5SO_4^{2-} + 20H^+ + 10e^-$

8. *Add & cancel*

   $2MnO_4^- + 2H_2O + 5SO_2 \longrightarrow 2Mn^{2+} + 5SO_4^{2-} + 4H^+$

Ion–Electron Method in Alkaline (Basic) Solution

1. Assign oxidation numbers to each atom in the equation.

2. Identify what is being oxidized and reduced.

3. Divide the skeleton equation into two half-reactions—one for the oxidation process and one for the reduction process.

4. For each half-reaction, balance atoms other than H and O.

5. For each half-reaction, balance oxygen by adding OH^- to the side that needs O. Balancing the Os will require adding twice as many OH^- as O needed. Balance hydrogen by adding H_2O to the side that needs H.

6. For each half-reaction, add up the total charge on each side and add electrons on either side as needed to balance the total charge on both sides of each half-reaction. (You should notice that the number of electrons corresponds to the change in oxidation state for each half-reaction.)

7. Balance the number of electrons being lost and gained.

8. Add the two half-reactions and cancel anything that is the same on both sides.

Example. $MnO_4^- + C_2O_4^{2-} \longrightarrow MnO_2 + CO_3^{2-}$ *under basic conditions*

1. *Oxidation #s* $(Mn^{7+}O^{2-}_4)^- + (C_2^{3+}O_4^{2-})^{2-} \longrightarrow Mn^{4+}O^{2-}_2 + (C^{4+}O^{2-}_3)^{2-}$

2. *Redox* Mn is being reduced and C is being oxidized.

3. *Half–reactions* $MnO_4^- \longrightarrow MnO_2$ *reduction*

   $C_2O_4^{2-} \longrightarrow CO_3^{2-}$ *oxidation*

4. *Balance atoms* The carbon atoms in the oxidation half-reaction need to be balanced.

   $C_2O_4^{2-} \longrightarrow 2CO_3^{2-}$

5. *Balance H & O* $MnO_4^- + 2H_2O \longrightarrow MnO_2 + 4OH^-$

   $C_2O_4^{2-} + 4OH^- \longrightarrow 2CO_3^{2-} + 2H_2O$

6. *Net charge* $MnO_4^- + 2H_2O \longrightarrow MnO_2 + 4OH^-$
   $\qquad\qquad\qquad -1 \qquad\qquad\qquad\qquad -4$
   $MnO_4^- + 2H_2O + 3e^- \longrightarrow MnO_2 + 4OH^-$

   $C_2O_4^{2-} + 4OH^- \longrightarrow 2CO_3^{2-} + 2H_2O$
   $\qquad\qquad -6 \qquad\qquad\quad -4$
   $C_2O_4^{2-} + 4OH^- \longrightarrow 2CO_3^{2-} + 2H_2O + 2e^-$

7. *Balance electrons* To balance electrons, the first half-reaction must be multiplied by 2 and the second half-reaction by 3.

   $2MnO_4^- + 4H_2O + 6e^- \longrightarrow 2MnO_2 + 8OH^-$
   $3C_2O_4^{2-} + 12OH^- \longrightarrow 6CO_3^{2-} + 6H_2O + 6e^-$

8. *Add & Cancel* $2MnO_4^- + 3C_2O_4^{2-} + 4OH^- \longrightarrow 2MnO_2 + 6CO_3^{2-} + 2H_2O$

Exercise 11–1: Balance the following redox equations using the oxidation states method.

1. ___Cu + ___AgNO$_3$ $\longrightarrow$ ___Ag + ___Cu(NO$_3$)$_2$

2. ___KBr + ___Fe$_2$(SO$_4$)$_3$ $\longrightarrow$ ___Br$_2$ + ___K$_2$SO$_4$ + ___FeSO$_4$

3. ___H$_2$SO$_3$ + ___I$_2$ + ___H$_2$O $\longrightarrow$ ___H$_2$SO$_4$ + ___HI

4. ___H$_2$O$_2$ + ___Ce(SO$_4$)$_2$ $\longrightarrow$ ___H$_2$SO$_4$ + ___O$_2$ + ___Ce$_2$(SO$_4$)$_3$

5. ___Bi(NO$_3$)$_3$ + ___Al + ___NaOH $\longrightarrow$ ___Bi + ___NH$_3$ + ___NaAlO$_2$

Exercise 11–2: Balance the following redox equations using the ion–electron method. Note that spectator ions have already been eliminated.

In acidic solutions:

1. ___IO_3^- + ___I^- + _____ $\longrightarrow$ ___I_2 + _____

2. ___MnO_4^- + ___$C_2O_4^{2-}$ + _____ $\longrightarrow$ ___Mn^{2+} + ___CO_2 + _____

3. ___CrO_4^{2-} + ___Cl^- + _____ $\longrightarrow$ ___Cr^{3+} + ___Cl_2 + _____

4. ___$Cr_2O_7^{2-}$ + ___S^{2-} + _____ $\longrightarrow$ ___Cr^{3+} + ___S + _____

In alkaline (basic) solutions:

5. ___Mn^{2+} + ___Br_2 + _____ $\longrightarrow$ ___MnO_2 + ___Br^- + _____

6. ___Al + ___MnO_4^- + _____ $\longrightarrow$ ___MnO_2 + ___$Al(OH)_4^-$ + _____

7. ___Mn^{2+} + ___ClO^- + _____ $\longrightarrow$ ___MnO_4^- + ___Cl^- + _____

8. ___CrO_2^- + ___ClO^- + _____ $\longrightarrow$ ___CrO_4^{2-} + ___Cl^- + _____

Predicting Types of Redox Reactions

A. Simple Redox

1. Hydrogen Displacement $Ca(s) + 2HOH \longrightarrow Ca(OH)_2(aq) + H_2(g)$
2. Metal Displacement $Zn(s) + CuSO_4(aq) \longrightarrow ZnSO_4(aq) + Cu(s)$
3. Halogen Displacement $Cl_2(g) + 2KBr(aq) \longrightarrow 2KCl(aq) + Br_2(aq)$
4. Combustion $CH_4(g) + 2O_2(g) \longrightarrow CO_2(g) + 2H_2O(l)$
5. Binary Compound Formation (also decomposition) $Cl_2(g) + 2Na(s) \longrightarrow 2NaCl(s)$

B. Reactions Involving Oxoanions such as $Cr_2O_7^{2-}$

$$14H^+(aq) + Cr_2O_7^{2-}(aq) + 6Fe^{2+}(aq) \longrightarrow 2Cr^{3+}(aq) + 7H_2O(l) + 6Fe^{3+}(aq)$$

C. Atypical Redox Reactions (those that appear to be ordinary single replacement reactions):

- Hydrogen reacts with a hot metallic oxide to produce the elemental metal and water.
- A metal sulfide reacts with oxygen to produce the metallic oxide and sulfur dioxide.
- Chlorine gas reacts with *dilute* sodium hydroxide to produce sodium hypochlorite, sodium chloride, and water.
- Copper reacts with *concentrated* sulfuric acid to produce copper(II) sulfate, sulfur dioxide, and water.
- Copper reacts with *dilute* nitric acid to produce copper(II) nitrate, nitrogen monoxide, and water.
- Copper reacts with *concentrated* nitric acid to produce copper(II) nitrate, nitrogen dioxide, and water.

To successfully predict the products of redox reactions, it must be remembered that as electrons are transferred, the oxidation numbers of at least two elements must change. All single replacement reactions, some synthesis and decomposition reactions, as well as all combustion reactions are redox reactions. It is important to be able to recognize not just the general types of redox reactions but also the atypical (tricky) ones listed above.

To determine if a reaction is a redox reaction, look at the reactants to see if there is both an oxidizing agent (a substance that tends to gain electrons) and a reducing agent (a substance that tends to lose electrons). If the problem states that the reaction occurs in either an acidic or basic solution, the reaction is most likely redox; hydrogen ions, hydroxide ions, and water should be included as appropriate. Spectator ions (those not involved in the redox process) need to be eliminated. The easiest way to eliminate most common spectator ions is to think about the most stable form of an element. For example, potassium ions will generally not be reduced to elemental potassium. The following list of common oxidizing and reducing agents and their products is very useful in predicting the products of redox reactions. *Note:* Use the reduction potential tables at the end of this chapter to determine appropriate reactants and products.

Common Oxidizing Agents	Products Formed
MnO_4^- in acidic solution	Mn^{2+}
MnO_2 in acidic solution	Mn^{2+}
MnO_4^- in neutral or basic solution	$MnO_2(s)$
$Cr_2O_7^{2-}$ in acidic solution	Cr^{3+}
HNO_3, concentrated	NO_2
HNO_3, dilute	NO
H_2SO_4, hot, concentrated	SO_2
metallic ions (higher oxidation #)	metallous ions (lower oxidation #)
free halogens	halide ions
Na_2O_2	$NaOH$
$HClO_4$	Cl^-
$C_2O_4^{2-}$	CO_2
H_2O_2	H_2O

Common Reducing Agents	Products Formed
halide ions	free halogen
free metals	metal ions
sulfite ions or SO_2	sulfate ions
nitrite ions	nitrate ions
free halogens, dilute basic solution	hypohalite ions
free halogens, concentrated basic solution	halate ions
metallous ions (lower oxidation #)	metallic ions (higher oxidation #)
H_2O_2	O_2

Example **A solution of tin(II) chloride is added to an acidified solution of potassium permanganate.**

$$5Sn^{2+} + 16H^+ + 2MnO_4^- \longrightarrow 5Sn^{4+} + 2Mn^{2+} + 8H_2O$$

Note: Since permanganate is reduced in acid solution to Mn^{2+}, it is logical for tin(II) to be oxidized. The word "acidified" indicates that hydrogen ions and water must be added. Chloride and potassium are spectator ions and a total of $10e^-$ are transferred.

Example **A solution of potassium iodide is added to an acidified solution of potassium dichromate.**

$$6I^- + 14H^+ + Cr_2O_7^{2-} \longrightarrow 2Cr^{3+} + 3I_2 + 7H_2O$$

Note: Since dichromate is reduced in acid solution to Cr^{3+}, it is logical for iodide to be oxidized to iodine. The word "acidified" indicates that hydrogen ions and water must be added. Potassium serves as a spectator ion and a total of $6e^-$ are transferred.

Example **Hydrogen peroxide is added to an acidified potassium bromide solution.**

$$H_2O_2 + 2H^+ + 2Br^- \longrightarrow Br_2 + 2H_2O$$

Note: Hydrogen peroxide is reduced to water and the bromide ions are oxidized to bromine. Potassium is a spectator ion and a total of $2e^-$ are transferred.

Equation Balancing Challenge: Balance the following equations.

For those who have mastered the balancing of any type of redox equation, an extra challenge is provided below. Balance *each* of the following equations. Show how your answers were obtained. *Note:* All of the reactions take place in aqueous solution.

Daughter of All Equations

$$HIO_3 + FeI_2 + HCl \longrightarrow FeCl_3 + ICl + HOH$$

Son of All Equations

$$CuSCN + KIO_3 + HCl \longrightarrow CuSO_4 + KCl + HCN + ICl + HOH$$

Father of All Equations

$$NH_4ClO_4 + Al \longrightarrow HCl + Cl_2 + NO + Al_2O_3 + HOH$$

Mother of All Equations

$$[Cr(N_2H_4CO)_6]_4[Cr(CN)_6]_3 + KMnO_4 + H_2SO_4 \longrightarrow$$

$$K_2Cr_2O_7 + MnSO_4 + CO_2 + KNO_3 + K_2SO_4 + HOH$$

These equations are not easy, but they all work! Have fun!

Exercise 11–3: Predict reactants and products for the following redox equations. Be sure to notice whether reactions occur in acidic or basic solution. Also, all spectator ions must be eliminated.

1. Potassium permanganate solution is added to concentrated hydrochloric acid.

2. Potassium dichromate solution is added to an acidified solution of sodium sulfite.

3. Solutions of potassium iodide, potassium iodate, and dilute sulfuric acid are mixed.

4. Manganese(IV) oxide is added to warm, concentrated hydrobromic acid.

5. Chlorine gas is bubbled into cold, dilute sodium hydroxide.

6. Hydrogen peroxide solution is added to acidified potassium iodide solution.

7. Hydrogen peroxide is added to an acidified solution of potassium dichromate.

8. Sulfur dioxide gas is bubbled through an acidified solution of potassium permanganate.

9. A solution of tin(II) ions is added to an acidified solution of potassium dichromate.

10. A small amount of solid iodine is shaken with 0.1 M sodium hydroxide.

Standard Electrode Potentials at 25 °C and 1 Atmosphere

Acid Solution

Half-Reaction			$E°$ (volts)
$F_2 + 2e^-$	$\longrightarrow$	$2F^-$	+2.87
$O_3 + 2H^+ + 2e^-$	$\longrightarrow$	$O_2 + HOH$	+2.07
$S_2O_8^{2-} + 2e^-$	$\longrightarrow$	$2SO_4^{2-}$	+2.01
$Co^{3+} + e^-$	$\longrightarrow$	Co^{2+}	+1.808
$H_2O_2 + 2H^+ + 2e^-$	$\longrightarrow$	$2HOH$	+1.776
$MnO_4^- + 4H^+ + 3e^-$	$\longrightarrow$	$MnO_2 + 2HOH$	+1.695
$Au^+ + e^-$	$\longrightarrow$	Au	+1.691
$PbO_2 + SO_4^{2-} + 4H^+ + 2e^-$	$\longrightarrow$	$PbSO_4 + 2HOH$	+1.682
$2HOCl + 2H^+ + 2e^-$	$\longrightarrow$	$Cl_2 + 2HOH$	+1.63
$Ce^{4+} + e^-$	$\longrightarrow$	Ce^{3+}	+1.61
$MnO_4^- + 8H^+ + 5e^-$	$\longrightarrow$	$Mn^{2+} + 4HOH$	+1.51
$Mn^{3+} + e^-$	$\longrightarrow$	Mn^{2+}	+1.51
$Au^{3+} + 3e^-$	$\longrightarrow$	Au	+1.498
$PbO_2 + 4H^+ + 2e^-$	$\longrightarrow$	$Pb^{2+} + 2HOH$	+1.455
$Au^{3+} + 2e^-$	$\longrightarrow$	Au^+	+1.402
$Cl_2(g) + 2e^-$	$\longrightarrow$	$2Cl^-$	+1.3595
$Cr_2O_7^{2-} + 14H^+ + 6e^-$	$\longrightarrow$	$2Cr^{3+} + 7HOH$	+1.33
$Tl^{3+} + 2e^-$	$\longrightarrow$	Tl^+	+1.25
$MnO_2 + 4H^+ + 2e^-$	$\longrightarrow$	$Mn^{2+} + 2HOH$	+1.23
$O_2 + 4H^+ + 4e^-$	$\longrightarrow$	$2HOH$	+1.229
$Br_2(l) + 2e^-$	$\longrightarrow$	$2Br^-$	+1.0652
$NO_3^- + 4H^+ + 3e^-$	$\longrightarrow$	$NO + 2HOH$	+0.96
$2Hg^{2+} + 2e^-$	$\longrightarrow$	Hg_2^{2+}	+0.920
$Hg^{2+} + 2e^-$	$\longrightarrow$	Hg	+0.854
$2NO_3^- + 4H^+ + 2e^-$	$\longrightarrow$	$N_2O_4 + 2HOH$	+0.803
$Ag^+ + e^-$	$\longrightarrow$	Ag	+0.7991
$Hg_2^{2+} + 2e^-$	$\longrightarrow$	$2Hg$	+0.788
$Fe^{3+} + e^-$	$\longrightarrow$	Fe^{2+}	+0.771
$O_2 + 2H^+ + 2e^-$	$\longrightarrow$	H_2O_2	+0.6824
$MnO_4^- + e^-$	$\longrightarrow$	MnO_4^{2-}	+0.564
$I_2 + 2e^-$	$\longrightarrow$	$2I^-$	+0.5355
$Cu^+ + e^-$	$\longrightarrow$	Cu	+0.521
$H_2SO_3 + 4H^+ + 4e^-$	$\longrightarrow$	$S + 3HOH$	+0.450

(continued on next page)

Standard Electrode Potentials at 25 °C and 1 Atmosphere

Acid Solution, continued

Half-Reaction			E° (volts)
$Cu^{2+} + 2e^-$	$\longrightarrow$	Cu	+0.337
$AgCl + e^-$	$\longrightarrow$	$Ag + Cl^-$	+0.2222
$SO_4^{2-} + 4H^+ + 2e^-$	$\longrightarrow$	$H_2SO_3 + HOH$	+0.172
$Sn^{4+} + 2e^-$	$\longrightarrow$	Sn^{2+}	+0.15
$S + 2H^+ + 2e^-$	$\longrightarrow$	$H_2S(aq)$	+0.142
$\mathbf{2H^+ + 2e^-}$	$\longrightarrow$	$\mathbf{H_2}$	**0.0000**
$Pb^{2+} + 2e^-$	$\longrightarrow$	Pb	–0.126
$Sn^{2+} + 2e^-$	$\longrightarrow$	Sn	–0.136
$Ni^{2+} + 2e^-$	$\longrightarrow$	Ni	–0.250
$H_3PO_4 + 2H^+ + 2e^-$	$\longrightarrow$	$H_3PO_3 + HOH$	–0.276
$Co^{2+} + 2e^-$	$\longrightarrow$	Co	–0.277
$Tl^+ + e^-$	$\longrightarrow$	Tl	–0.3363
$PbSO_4 + 2e$	$\longrightarrow$	$Pb + SO_4^{2-}$	–0.3588
$Cd^{2+} + 2e^-$	$\longrightarrow$	Cd	–0.4029
$Cr^{3+} + e^-$	$\longrightarrow$	Cr^{2+}	–0.408
$Fe^{2+} + 2e^-$	$\longrightarrow$	Fe	–0.4402
$Ga^{3+} + 3e^-$	$\longrightarrow$	Ga	–0.529
$Cr^{3+} + 3e^-$	$\longrightarrow$	Cr	–0.744
$Zn^{2+} + 2e^-$	$\longrightarrow$	Zn	–0.7628
$Mn^{2+} + 2e^-$	$\longrightarrow$	Mn	–1.180
$Al^{3+} + 3e^-$	$\longrightarrow$	Al	–1.662
$Be^{2+} + 2e^-$	$\longrightarrow$	Be	–1.847
$Mg^{2+} + 2e^-$	$\longrightarrow$	Mg	–2.363
$Ce^{3+} + 3e^-$	$\longrightarrow$	Ce	–2.483
$Na^+ + e^-$	$\longrightarrow$	Na	–2.714
$Ca^{2+} + 2e^-$	$\longrightarrow$	Ca	–2.866
$Sr^{2+} + 2e^-$	$\longrightarrow$	Sr	–2.888
$Ba^{2+} + 2e^-$	$\longrightarrow$	Ba	–2.906
$Ra^{2+} + 2e^-$	$\longrightarrow$	Ra	–2.916
$Cs^+ + e^-$	$\longrightarrow$	Cs	–2.923
$Rb^+ + e^-$	$\longrightarrow$	Rb	–2.925
$K^+ + e^-$	$\longrightarrow$	K	–2.925
$Li^+ + e^-$	$\longrightarrow$	Li	–3.045

Standard Electrode Potentials at 25 °C and 1 Atmosphere

Alkaline Solution

Half-Reaction			$E°$ (volts)
$HO_2^- + HOH + 2e^-$	$\longrightarrow$	$3OH^-$	+0.878
$NiO_2 + 2HOH + 2e^-$	$\longrightarrow$	$Ni(OH)_2 + 2OH^-$	+0.490
$O_2 + 2HOH + 4e^-$	$\longrightarrow$	$4OH^-$	+0.401
$NO_3^- + HOH + 2e^-$	$\longrightarrow$	$NO_2^- + 2OH^-$	+0.01
$CrO_4^{2-} + 4HOH + 3e^-$	$\longrightarrow$	$Cr(OH)_3 + 5OH^-$	−0.13
$S + 2e^-$	$\longrightarrow$	S^{2-}	−0.447
$Cd(OH)_2 + 2e^-$	$\longrightarrow$	$Cd + 2OH^-$	−0.809
$2HOH + 2e^-$	$\longrightarrow$	$H_2 + 2OH^-$	−0.82806
$Fe(OH)_2 + 2e^-$	$\longrightarrow$	$Fe + 2OH^-$	−0.877
$Zn(OH)_4^{2-} + 2e^-$	$\longrightarrow$	$Zn + 4OH^-$	−1.215
$Al(OH)_4^- + 3e^-$	$\longrightarrow$	$Al + 4OH^-$	−2.33

Data for Standard Electrode Potentials from A. J. de Bethune and N. A. Swendeman Loud, "Table of Electrode Potentials and Temperature Coefficients," pages 414–424 in *Encyclopedia of Electrochemistry* (C. A. Hampel, editor), Van Nostrand Reinhold, New York, 1964, and from A. J. de Bethune and N. A. Swendman Loud, *Standard Aqueous Electrode Potentials and Temperature Coefficients*, 19 pages, C. A. Hampel publisher, Skokie, Illinois, 1964.

Notes

Chapter 12
Electrolysis in Aqueous Solutions

Electrolysis of salts in aqueous solutions is a complex process. In theory, E^o values (Standard Reduction Potentials) can be used to predict which element will plate out at a particular electrode when various solutions are combined. For example, suppose an electrolytic cell contains the ions Cu^{2+}, Ag^+, and Zn^{2+}. If the voltage is initially very low and is gradually turned up, which metal will plate out first onto the cathode? Which will plate out second? A look at the standard reduction potentials for these ions reveals the following values:

$$Ag^+ + e^- \longrightarrow Ag \qquad\qquad E^o = +0.80 \text{ V}$$

$$Cu^{2+} + 2e^- \longrightarrow Cu \qquad\qquad E^o = +0.34 \text{ V}$$

$$Zn^{2+} + 2e^- \longrightarrow Zn \qquad\qquad E^o = -0.76 \text{ V}$$

Remember that the more *positive* the E^o value, the more the reaction has a tendency to proceed in the direction indicated. Of the three reactions listed, the reduction of Ag^+ occurs most easily, and the ease of reduction is as follows:

$$Ag^+ > Cu^{2+} > Zn^{2+}$$

As the voltage is increased, silver will plate out first, followed by copper, and then zinc.

Standard reduction potentials can be used to calculate voltages for both electrolytic and voltaic cells. The minimum applied voltage necessary to bring about a nonspontaneous oxidation–reduction reaction in an electrolytic cell under standard conditions can be calculated by adding the proper potentials. In laboratory practice, however, it is found that the voltage required to operate an electrolytic cell is somewhat higher than that calculated from electrode potentials. The excess voltage, referred to as *overvoltage*, may be 1 V or more; it is particularly large when one of the products of the cell reaction is a gas such as hydrogen or oxygen. Although the mechanism of overvoltage is poorly understood, it is known to arise from kinetic effects. Electrode processes involving the transfer of electrons between a metal electrode and ions in solution tend to be slow. Applying a voltage higher than that theoretically required supplies the activation energy necessary to make the reaction proceed at a finite rate.

Caution: It is risky using only $E°$ values to predict the products of an electrolysis reaction in water!

Some simple rules help to predict aqueous electrolysis equations on the AP Chemistry Exam. However, the presence of water does increase the number of possible reactions that can take place in an electrolytic cell. We must first consider what are the possible electrode reactions and then consider some specific examples.

Anode—electrode where oxidation occurs!

Cathode—electrode where reduction occurs!

Rules for Predicting Cathode Reactions (Reduction)

When a direct electric current is passed through a water solution of an electrolyte, two possible reduction processes may occur at the cathode.

- The cation may be reduced to the corresponding metal.

$$M^{n+} + ne^- \longrightarrow M(s)$$

$$(n = \text{charge of cation})$$

- Water molecules may be reduced to elementary hydrogen.

$$2HOH + 2e^- \longrightarrow H_2(g) + 2OH^-(aq)$$

The relative ease of reduction of the cation as opposed to a water molecule is important in determining which of these two reactions will occur.

For salts containing transition metal cations (e.g., Cu^{2+}, Ag^+, Ni^{2+}), which are relatively easy to reduce compared to water, Reaction #1 will occur at the cathode and the transition metal will plate out.

On the other hand, if the cation is a representative metal (e.g., Na^+, Mg^{2+}, Al^{3+}), the water molecules will be easier to reduce compared to the cation, and Reaction #2 will occur at the cathode, producing hydrogen gas and hydroxide ions.

Rules for Predicting Anode Reactions (Oxidation)

The oxidation process that occurs at the anode of an electrolytic cell operating in aqueous solution may be one of two oxidation processes.

- The anion may be oxidized to the corresponding nonmetal.

$$2X^- \longrightarrow X_2 + 2e^-$$

- Water molecules may be oxidized to elementary oxygen.

$$HOH \longrightarrow \tfrac{1}{2}O_2(g) + 2H^+(aq) + 2e^-$$

Again, the relative ease of oxidation of the anion compared to that of a water molecule is the major factor in determining which of these two reactions will take place.

For salts containing iodide, bromide, or chloride ions, it is usually easier to oxidize these nonmetals rather than water. It will be found that the nonmetal (I_2, Br_2, Cl_2) is formed at the anode.

When the anion present is any other ion that is more difficult to oxidize than water (e.g., F^-, SO_4^{2-}, etc.), Reaction #2 will occur at the anode, producing elementary oxygen and hydrogen (hydronium) ions.

Based on their oxidation potentials, water should be more readily oxidized than chloride ions.

$$2Cl^- \longrightarrow Cl_2 + 2e^- \qquad\qquad E° = -1.36 \text{ V}$$

$$2HOH \longrightarrow O_2 + 4H^+ + 4e^- \qquad\qquad E° = -1.23 \text{ V}$$

Water has a more positive potential and we should expect to see O_2 produced at the anode because it is thermodynamically easier to oxidize HOH than Cl^-. In reality Cl_2 gas is produced rather than O_2. The overvoltage is much greater for the production of O_2 than for Cl_2, which explains why the chlorine is produced first.

Sample Electrolysis Reactions (using the rules from page 82)

1. Copper(II) chloride in water

$$Cu^{2+}(aq) + 2Cl^-(aq) \longrightarrow Cu(s) + Cl_2(g)$$

2. Copper(II) sulfate in water

$$Cu^{2+}(aq) + HOH(l) \longrightarrow Cu(s) + \tfrac{1}{2}O_2(g) + 2H^+(aq)$$

3. Sodium chloride in water

$$2Cl^-(aq) + 2HOH(l) \longrightarrow H_2(g) + Cl_2(g) + 2OH^-(aq)$$

4. Sodium sulfate in water

$$2HOH(l) \longrightarrow 2H_2(g) + O_2(g)$$

Please note that all electrolysis reactions that may be included in the Equations Section on the AP Chemistry Exam assume that inert electrodes are being used. Platinum electrodes are inert and do not appear in the equation. Also assume that carbon (graphite) electrodes are inert. Graphite does react with oxygen gas at high temperatures to form carbon dioxide, but this problem is never encountered on the AP Chemistry Exam. If an anode is made of aluminum, the Al can combine with the oxygen gas being produced to form aluminum oxide. This process is usually referred to as anodizing. Anodizing has **never** been included in the Equation Section of the AP Chemistry Exam.

In contrast to the electrolysis of aqueous salts, the electrolysis of molten binary salts is much simpler and more straightforward. The anion in the salt is oxidized at the anode and the cation is reduced at the cathode.

Electrolysis of molten sodium chloride:

$$2NaCl(l) \longrightarrow 2Na(l) + Cl_2(g)$$

Remember that in aqueous solution the electrolysis of sodium chloride results in the following reaction:

$$2Na^+(aq) + 2Cl^-(aq) + 2HOH(l) \longrightarrow H_2(g) + Cl_2(g) + 2Na^+(aq) + 2OH^-(aq)$$

Fortunately for the AP Chemistry student, the electrolysis of molten ternary salts is not studied in introductory college or general chemistry classes. Life would be a lot more complex if one had to deal with the electrolysis of molten sodium nitrate.

Exercise 12–1: Write balanced net ionic equations for electrolysis of the following substances using platinum electrodes.

1. aqueous potassium fluoride

2. aqueous nickel(II) nitrate

3. molten aluminum oxide

4. aqueous cesium bromide

5. aqueous chromium(III) iodide

6. aqueous magnesium sulfide

7. aqueous ammonium chloride

8. molten lithium fluoride

9. aqueous gold(III) acetate

10. aqueous cobalt(II) bromide

Chapter 13
Complex Ion Reactions

Coordination Chemistry

Reactions involving complex ions and coordination chemistry are found rather sparingly on the reaction prediction section of the AP exam. However, some knowledge of general complex-ion reactions can be helpful in predicting products. Also, a number of complex-ion compounds are used in the qualitative analysis of metal ion solutions. Because many of these complex ions have unique coloration and can be easily formed or dissolved through addition of acid and base, they provide a rapid means of distinguishing one metal ion from another. Review the section in Chapter 5 on naming complex ions before beginning this chapter.

Since complex ions generally involve transition metals such as iron, cobalt, nickel, chromium, copper, zinc and silver, in addition to aluminum, the presence of these metals among the reactants can signal a complex ion reaction. Also, many complex ion reactions occur with the addition of an excess of a substance containing the ligand. Therefore, wording such as "an excess of concentrated ammonia solution," "an excess of sodium or potassium hydroxide solution," or "an excess of potassium cyanide or ammonium thiocyanate solution," are all clues to a possible complex ion reaction. Finally, many complex ion structures will break down when there is an addition of a concentrated acid solution.

Common complex ion metals (Lewis acids):   **Fe, Co, Ni, Cr, Cu, Zn, Ag, Al**

Common ligands (Lewis bases): NH_3, CN^-, OH^-, SCN^-

> General Rule: The number of ligands will often be twice the charge of the metal ion.

Types of Complex Ion Reactions

Note: For each reaction shown below, net ionic equations have been written with the elimination of spectator ions.

Formation of ammine complexes

Metal ion solutions react with an excess of concentrated ammonia to form ammine complex ions.

Copper(II) chloride solution is combined with an excess of concentrated ammonia solution.

$$Cu^{2+}(aq) + 4NH_3(aq) \longrightarrow [Cu(NH_3)_4]^{2+}(aq)$$

Solid metallic hydroxides, when combined with concentrated ammonia solution, produce ammine complex ions and hydroxide ions.

An excess of concentrated ammonia solution is added to freshly precipitated copper(II) hydroxide.

$$4NH_3(aq) + Cu(OH)_2(s) \longrightarrow [Cu(NH_3)_4]^{2+}(aq) + 2OH^-(aq)$$

Formation of cyanide complexes

Metal ion solutions react with an excess of cyanide solution to form cyano complex ions.

Excess sodium cyanide solution is added to a solution of silver nitrate.

$$2CN^-(aq) + Ag^+(aq) \longrightarrow [Ag(CN)_2]^-(aq)$$

Formation of hydroxo complexes

Metal ion solutions react with an excess of hydroxide solution to form hydroxo complex ions.

Excess potassium hydroxide solution is added to a solution of aluminum nitrate.

$$4OH^-(aq) + Al^{3+}(aq) \longrightarrow [Al(OH)_4]^-(aq) \quad or \quad Al(OH)_3(s)$$

Solid metallic hydroxides, when added to an excess of hydroxide solution, produce hydroxo complex ions.

An excess of sodium hydroxide solution is added to a precipitate of aluminum hydroxide in water.

$$OH^-(aq) + Al(OH)_3(s) \longrightarrow [Al(OH)_4]^-(aq)$$

Formation of thiocyanato complexes

Metal ion solutions react with thiocyanate solution to form thiocyanato complex ions.

A solution of ammonium thiocyanate is added to a solution of iron(III) chloride

$$SCN^-(aq) + Fe^{3+}(aq) \longrightarrow [Fe(SCN)]^{2+}(aq) \quad or \quad [Fe(SCN)_6]^{3-}(aq)$$

Dissolution of a complex ion with acid

A complex ion solution is treated with a strong acid solution resulting in the free metal ion or a precipitate of a metal salt.

A solution of diamminesilver(I) chloride is treated with dilute nitric acid.

$$[Ag(NH_3)_2]^+(aq) + Cl^-(aq) + 2H^+(aq) \longrightarrow AgCl(s) + 2NH_4^+(aq)$$

Excess dilute nitric acid is added to a solution containing the tetraamminecadmium(II) ion.

$$4H^+(aq) + [Cd(NH_3)_4]^{2+}(aq) \longrightarrow Cd^{2+}(aq) + 4NH_4^+(aq)$$

Lewis acid–base reaction

An electron pair acceptor (Lewis acid) is combined with an electron pair donor (Lewis base) to form a coordinate-covalent compound.

The gases boron trifluoride and ammonia are mixed.

$$BF_3(g) + NH_3(g) \longrightarrow H_3N{:}BF_3$$

Note: In the reaction above, BF_3 acts as the Lewis acid and NH_3 as the Lewis base.

Exercise 13–1: Predict products and write balanced net ionic equations for the following reactions.

1. Concentrated ammonia solution is added in excess to a solution of copper(II) nitrate.

2. Dilute hydrochloric acid solution is added to a solution of diamminesilver(I) nitrate.

3. An excess of nitric acid solution is added to a solution of tetraamminecopper(II) sulfate.

4. A suspension of zinc hydroxide is treated with concentrated sodium hydroxide solution.

5. A suspension of copper(II) hydroxide is treated with an excess of ammonia water.

6. A drop of potassium thiocyanate solution is added to a solution of iron(III) chloride.

7. Silver chloride is dissolved in excess ammonia solution.

8. Solid aluminum oxide is added to a solution of sodium hydroxide.

9. A concentrated solution of ammonia is added to a solution of zinc iodide.

10. An excess of sodium hydroxide solution is added to a solution of aluminum chloride.

Some possible questions for complex ion reactions:

- Identify evidence of a chemical change.
- Identify changes in color between reactants and products.
- Identify any spectator ions.
- Name the product.
- Describe the effect of adding acid.

Example   An excess of concentrated ammonia solution is added to freshly precipitated copper(II) hydroxide.

What evidence of a chemical reaction will be observed?

Solution $4NH_3(aq) + Cu(OH)_2(s) \longrightarrow [Cu(NH_3)_4]^{2+}(aq) + 2OH^-(aq)$

The solid copper(II) hydroxide will dissolve as the aqueous dark blue copper complex ion is formed.

Example   Copper(II) chloride solution is combined with an excess of concentrated ammonia solution.

Name the product(s) of the chemical reaction.

Solution $Cu^{2+}(aq) + 4NH_3(aq) \longrightarrow [Cu(NH_3)_4]^{2+}(aq)$

The product is the tetraamminecopper(II) ion.   (Chloride acts as a spectator ion.)

Example   A solution of diamminesilver(I) chloride is treated with dilute nitric acid.

If the product(s) of this reaction are filtered, what will be observed?

Solution $[Ag(NH_3)_2]^+(aq) + Cl^-(aq) + 2H^+(aq) \longrightarrow AgCl(s) + 2NH_4^+(aq)$

Solid white silver chloride will be collected on the filter paper.

Example   An excess of sodium hydroxide solution is added to a precipitate of aluminum hydroxide in water.

Identify the Lewis base in this reaction.

Solution $Al(OH)_3(s) + OH^-(aq) \longrightarrow [Al(OH)_4]^-(aq)$

The hydroxide ion acts as a Lewis base in this reaction by donating a pair of electrons to aluminum.

Chapter 14
Summary of Reactions

Anhydrides

- Metallic hydrides plus water produce hydrogen gas and metallic hydroxides.

- Soluble metallic oxides and water form bases (metallic hydroxides).

- Group 1 and 2 metallic nitrides react with water to produce metallic hydroxides and ammonia.

- Soluble nonmetallic oxides and water form acids. *Note:* The nonmetal retains its oxidation number.

Combustion

- Hydrocarbons and other organic compounds combine with excess oxygen to form carbon dioxide and water.

- Metals combine with oxygen to form metallic oxides.

- Binary compounds of nonmetals with hydrogen combine with oxygen to form water and nonmetal oxides.

- Nonmetallic sulfides combine with oxygen to form sulfur dioxide and nonmetal oxides.

Complex Ions

- Complex ion solutions treated with a strong acid solution produce the free metal ion or a precipitate of the metal salt with the ligand ion.

- An electron pair acceptor is combined with an electron pair donor to form a coordinate covalent compound.

- Metal ion solutions react with an excess of concentrated ammonia to form ammine complex ions.

- Metal ion solutions react with an excess of cyanide solution to form cyano complex ions.

- Metal ion solutions react with an excess of hydroxide solution to form hydroxo complex ions.

- Metal ion solutions react with thiocyanate solution to form thiocyanato complex ions.

- Solid metallic hydroxides when combined with concentrated ammonia solution produce soluble ammine complex ions and hydroxide ions.

- Solid metallic hydroxides when added to hydroxide solution produce hydroxo complex ions.

Decomposition

- Ammonium carbonate decomposes into ammonia, water, and carbon dioxide.

- Ammonium hydroxide decomposes into ammonia and water.

- Binary ionic compounds (molten) can be electrolyzed into their metal and nonmetal components.

- Carbonic acid decomposes into water and carbon dioxide.

- Hydrogen peroxide decomposes into water and oxygen.

- Metallic carbonates decompose into metallic oxides and carbon dioxide.

- Metallic chlorates decompose into metallic chlorides and oxygen.

- Oxyacids decompose into water and a nonmetallic oxide.

- Sulfurous acid decomposes into water and sulfur dioxide.

Synthesis

- A binary molecular compound combined with a nonmetal (contained in the compound) forms a single compound.

- An electron pair acceptor is combined with an electron pair donor to form a coordinate covalent compound.

- A halogen is added to an alkene forming a halogenated alkane.

- Hydrogen is added to an alkene forming an alkane.

- Metals and nonmetals combine to form binary ionic compounds.

- Metal ion solutions react with an excess of concentrated ammonia to form ammine complex ions.

- Metal ion solutions react with an excess of cyanide solution to form cyano complex ions.

- Metal ion solutions react with an excess of hydroxide solution to form hydroxo complex ions.

- Metal oxides combine with carbon dioxide to form metallic carbonates.

- Metal oxides combine with sulfur dioxide to form metallic sulfites.

- Nonmetallic oxides and water form acids. *Note:* The nonmetal retains its oxidation number.

- Soluble metallic oxides and water form bases (metallic hydroxides).

Metathesis (Double Replacement)

• Two soluble ions in aqueous solution may form an insoluble precipitate.

• Metal sulfides when combined with any acid will form hydrogen sulfide gas and a salt.

• Metallic carbonates when combined with an acid will form carbon dioxide gas, water, and a salt.

• Metallic sulfites when combined with an acid will form sulfur dioxide gas, water, and a salt.

• Ammonium salts when heated with a soluble strong hydroxide will form ammonia gas, water, and a salt.

• An acid and a base will form a salt and water.

• A salt formed from a strong acid and a weak base will hydrolyze in water to form a strong acid and a weak base.

• A salt formed from a weak acid and a strong base will hydrolyze in water to form a weak acid and a strong base.

Redox

• Binary ionic compounds (molten) can be electrolyzed into their metal and nonmetal components.

• Chlorine gas reacts with *dilute* sodium hydroxide to produce sodium hypochlorite, sodium chloride, and water.

• Copper reacts with *concentrated* nitric acid to produce copper(II) nitrate, nitrogen dioxide, and water.

• Copper reacts with *dilute* nitric acid to produce copper(II) nitrate, nitrogen monoxide, and water.

• Copper reacts with *concentrated* sulfuric acid to produce copper(II) sulfate, sulfur dioxide, and water.

- A halogen is added to an alkane forming a halogenated alkane.

- A halogen is added to an alkene forming a halogenated alkane.

- Active free halogens replace less active halide ions from their compounds in aqueous solution to form a halogen and halide ion in solution.

- Hydrocarbons and other organic compounds combine with excess oxygen to form carbon dioxide and water.

- Hydrogen gas is added to an alkene forming an alkane.

- Hydrogen gas reacts with a hot metallic oxide to produce the elemental metal and water.

- Metals and nonmetals can combine to form binary ionic compounds.

- Active free metals replace hydrogen in acids to form metallic ions and hydrogen gas.

- Active free metals replace hydrogen in water to form metallic hydroxides and hydrogen gas.

- Active free metals replace less active metals from their compounds in aqueous solution to form a metal and metal ion in solution.

- Metal sulfides react with oxygen to produce metallic oxides and sulfur dioxide.

- Binary compounds of nonmetals with hydrogen combine with oxygen to form nonmetal oxides and water.

- Nonmetallic sulfides combine with oxygen to form nonmetal oxides and sulfur dioxide.

Single Replacement

- Active free halogens replace less active halide ions from their compounds in aqueous solution to form a halogen and halide ion in solution.

- Active free metals replace hydrogen in acids to form metallic ions and hydrogen gas.

- Active free metals replace hydrogen in water to form metallic hydroxides and hydrogen gas.

- Active free metals replace less active metals from their compounds in aqueous solution to form a metal and metal ion in solution.

Atypical Redox Reactions

Note: The following reactions look like single replacements from their reactants but are actually tricky redox reactions.

- Hydrogen reacts with a hot metallic oxide to produce the elemental metal and water.

- Metal sulfides react with oxygen to produce metallic oxides and sulfur dioxide.

- Chlorine gas reacts with *dilute* sodium hydroxide to produce sodium hypochlorite, sodium chloride, and water.

- Copper reacts with *concentrated* sulfuric acid to produce copper(II) sulfate, sulfur dioxide, and water.

- Copper reacts with *dilute* nitric acid to produce copper(II) nitrate, nitrogen monoxide, and water.

- Copper reacts with *concentrated* nitric acid to produce copper(II) nitrate, nitrogen dioxide, and water.

Chapter 15
Molecular Equations for Reference

Aluminum

$11Al + 3BiONO_3 + 11KOH \longrightarrow 3Bi + 3NH_3 + 11KAlO_2 + HOH$

$2Al + 3CdCl_2 \longrightarrow 2AlCl_3 + 3Cd$

$Al + GaCl_3 \longrightarrow AlCl_3 + Ga$

$4Al + 3GeO_2 \longrightarrow 2Al_2O_3 + 3Ge$

$2Al + 2HOH + 2NaOH \longrightarrow 2NaAlO_2 + 3H_2$

$8Al + 3Pb_3O_4 \longrightarrow 4Al_2O_3 + 9Pb$

$4Al + 3KSCN + 18HCl \longrightarrow 3KCl + 4AlCl_3 + 3NH_4Cl + 3C + 3H_2S$

$Al(OH)_3 + 3HNO_3 \longrightarrow Al(NO_3)_3 + 3HOH$

$Al(OH)_3 + NaOH \longrightarrow Na[Al(OH)_4]$

$2AlK(SO_4)_2 + 6NH_4OH \longrightarrow 2Al(OH)_3 + 3(NH_4)_2SO_4 + K_2SO_4$

Arsenic

$5As_2O_3 + 4HIO_3 + 13HOH \longrightarrow 10H_3AsO_4 + 2I_2$

$As_2O_3 + 2HNO_3 \longrightarrow As_2O_5 + HOH + NO_2 + NO$

$As_2O_3 + 2I_2 + 10NaHCO_3 \longrightarrow 2Na_3AsO_4 + 4NaI + 10CO_2 + 5HOH$

$3As_2S_5 + 9HOH + 5HClO_3 \longrightarrow 5HCl + 6H_3AsO_4 + 15S$

$As_4O_6 + 12HCl \longrightarrow 4AsCl_3 + 6HOH$

$As_4O_6 + 12NaOH \longrightarrow 4Na_3AsO_3 + 6HOH$

Barium

$BaCO_3 + K_2Cr_2O_7 \longrightarrow K_2CrO_4 + BaCrO_4 + CO_2$

$BaC_2 + 2HOH \longrightarrow C_2H_2 + Ba(OH)_2$

$2BaCl_2 + 2NaHCO_3 \longrightarrow 2BaCO_3 + 2NaCl + 2HCl$

$Ba(ClO_3)_2 \longrightarrow BaCl_2 + 3O_2$

$2BaCrO_4 + 6KI + 16HCl \longrightarrow 2BaCl_2 + 2CrCl_3 + 6KCl + 8HOH + 3I_2$

$BaHPO_4 + 2HCl \longrightarrow BaCl_2 + H_3PO_4$

$2Ba(NO_3)_2 \longrightarrow 2BaO + 4NO_2 + O_2$

$BaO + CO_2 \longrightarrow BaCO_3$

$BaO + 2HNO_3 \longrightarrow Ba(NO_3)_2 + HOH$

$BaO + HOH \longrightarrow Ba(OH)_2$

$2BaO + O_2 \longrightarrow 2BaO_2$

$2BaO + 2SO_2 \longrightarrow 2BaSO_3$

$Ba(OH)_2 + Ca(HCO_3)_2 \longrightarrow BaCO_3 + CaCO_3 + 2HOH$

$BaS + 2O_2 \longrightarrow BaSO_4$

$BaSO_4 + NaKCO_3 \longrightarrow NaKSO_4 + BaCO_3$

Beryllium

$Be + Br_2 \longrightarrow BeBr_2$

Bismuth

$4Bi + 3O_2 \longrightarrow 2Bi_2O_3$

Boron

$$BF_3 + NH_3 \longrightarrow H_3N{:}BF_3$$
$$BF_3 + NaF \longrightarrow Na[BF_4]$$

Bromine

$$3Br_2 + 2Sb \longrightarrow 2SbBr_3$$
$$2Br_2 + C_2H_2 \longrightarrow C_2H_2Br_4$$
$$Br_2 + C_6H_6 \longrightarrow C_6H_5Br + HBr$$
$$Br_2 + 2HI \longrightarrow 2HBr + I_2$$
$$6Br_2 + 10KClO_3 + 5H_2SO_4 + 6HOH \longrightarrow 12HBrO_3 + 5K_2SO_4 + 10HCl$$

Cadmium

$$2Cd + O_2 \longrightarrow 2CdO$$
$$Cd(OH)_2 + 4NH_3 \longrightarrow [Cd(NH_3)_4](OH)_2$$

Calcium

$$Ca + 2HOH \longrightarrow Ca(OH)_2 + H_2$$
$$CaBr_2 + MnO_2 + 3H_2SO_4 \longrightarrow Ca(HSO_4)_2 + MnSO_4 + 2HOH + Br_2$$
$$CaCO_3 + CO_2 + HOH \longrightarrow Ca(HCO_3)_2$$
$$CaCO_3 + 2HBr \longrightarrow CaBr_2 + CO_2 + HOH$$
$$CaCO_3 + 2HCl \longrightarrow CaCl_2 + CO_2 + HOH$$
$$CaC_2 + N_2 \longrightarrow Ca(CN)_2$$
$$5CaC_2O_4 + 8H_2SO_4 + 2KMnO_4 \longrightarrow K_2SO_4 + 2MnSO_4 + 5CaSO_4 + 10CO_2 + 8HOH$$
$$Ca(ClO)_2 + As_2O_3 \longrightarrow As_2O_5 + CaCl_2$$
$$CaHPO_4 + 2HCl \longrightarrow CaCl_2 + H_3PO_4$$
$$Ca(HCO_3)_2 + 2NaOH \longrightarrow CaCO_3 + Na_2CO_3 + 2HOH$$
$$CaO + 2HCl \longrightarrow CaCl_2 + HOH$$
$$CaO + H_2S \longrightarrow CaS + HOH$$
$$CaO + SO_3 \longrightarrow CaSO_4$$
$$Ca(OH)_2 + CO_2 \longrightarrow CaCO_3 + HOH$$
$$Ca(OH)_2 + H_2O_2 \longrightarrow CaO_2 + 2HOH$$
$$Ca(OH)_2 + 2NH_4NO_3 \longrightarrow Ca(NO_3)_2 + 2NH_3 + 2HOH$$
$$Ca_3N_2 + 6HOH \longrightarrow 3Ca(OH)_2 + 2NH_3$$
$$Ca_3P_2 + 3HOH \longrightarrow 3CaO + 2PH_3$$

Carbon

$$C + O_2 \longrightarrow CO_2$$
$$2C + O_2 \longrightarrow 2CO$$
$$CH_4 + I_2 \longrightarrow CH_3I + HI$$
$$2C_2H_6 + 7O_2 \longrightarrow 4CO_2 + 6HOH$$
$$CH_3Cl + KOH \longrightarrow KCl + CH_3OH$$
$$CH_3COOH + CH_3OH \longrightarrow CH_3COOCH_3 + HOH$$
$$2C_2H_5OH \longrightarrow (C_2H_5)_2O + HOH$$
$$CO_2 + 2NaOH \longrightarrow Na_2CO_3 + HOH$$

Cesium

$$2Cs + Cl_2 \longrightarrow 2CsCl$$
$$2Cs + 2HOH \longrightarrow 2CsOH + H_2$$
$$2CsMnO_4 + 9H_2SO_4 + 5CaC_2O_4 \longrightarrow 5CaSO_4 + 2CsHSO_4 + 2MnSO_4 + 10CO_2 + 8HOH$$

Chlorine

$3Cl_2 + 2Sb \longrightarrow 2SbCl_3$

$Cl_2 + SbCl_3 \longrightarrow SbCl_5$

$12Cl_2 + 4AsH_3 + 6HOH \longrightarrow As_4O_6 + 24HCl$

$14Cl_2 + As_2S_3 + 20HOH \longrightarrow 2H_3AsO_4 + 3H_2SO_4 + 28HCl$

$Cl_2 + CH_2Cl_2 \longrightarrow HCl + CHCl_3$

$3Cl_2 + 2CrCl_3 + 16NaCH_3COO + 8HOH \longrightarrow 12NaCl + 2Na_2CrO_4 + 16CH_3COOH$

$27Cl_2 + 2CrI_3 + 64KOH \longrightarrow 2K_2CrO_4 + 6KIO_4 + 54KCl + 32HOH$

$Cl_2 + HOH \longrightarrow HCl + HClO$

$Cl_2 + HOH + H_2SO_3 \longrightarrow H_2SO_4 + 2HCl$

$Cl_2 + 2KI \longrightarrow 2KCl + I_2$

$Cl_2 + 2NaOH \longrightarrow NaOCl + NaCl + HOH$

$3Cl_2 + 6NaOH \longrightarrow 5NaCl + NaClO_3 + 3HOH$

$Cl_2O_3 + HOH \longrightarrow 2HClO_2$

Chromium

$2CrO_3 + 6HCl + 3C_2H_5OH \longrightarrow 2CrCl_3 + 3CH_3CHO + 6HOH$

$2CrO_3 + 6FeSO_4 + 6H_2SO_4 \longrightarrow Cr_2(SO_4)_3 + 3Fe_2(SO_4)_3 + 6HOH$

Cobalt

$3Co(CH_3COO)_2 + 2K_3[Fe(CN)_6] \longrightarrow Co_3[Fe(CN)_6]_2 + 6KCH_3COO$

$6CoCl_2 + 5HgO + 2KMnO_4 + 9HOH \longrightarrow 6Co(OH)_3 + 5HgCl_2 + 2KCl + 2MnO_2$

$Co(OH)_2 + 6NH_3 \longrightarrow [Co(NH_3)_6](OH)_2$

$3CoS + 8HNO_3 \longrightarrow 2NO + 4HOH + 3S + 3Co(NO_3)_2$

Copper

$3Cu + 8HNO_3 \longrightarrow 3Cu(NO_3)_2 + 2NO + 4HOH$

$Cu + 2H_2SO_4 \longrightarrow CuSO_4 + 2HOH + SO_2$

$Cu + 2FeCl_3 \longrightarrow CuCl_2 + 2FeCl_2$

$[Cu(NH_3)_4]SO_4 + 4HNO_3 + 4HOH \longrightarrow [Cu(H_2O)_4]SO_4 + 4NH_4NO_3$

$2Cu(NO_3)_2 + 4KI \longrightarrow 2CuI + 4KNO_3 + I_2$

$CuO + H_2 \longrightarrow HOH + Cu$

$4CuSCN + 7KIO_3 + 14HCl \longrightarrow 4CuSO_4 + 7ICl + 4HCN + 7KCl + 5HOH$

$CuSO_4 + 6NH_4OH \longrightarrow [Cu(NH_3)_6]SO_4 + 6HOH$

$2CuSO_4 + 2KI + 2FeSO_4 \longrightarrow 2CuI + K_2SO_4 + Fe_2(SO_4)_3$

$2CuSO_4 + K_4[Fe(CN)_6] \longrightarrow 2K_2SO_4 + Cu_2[Fe(CN)_6]$

Fluorine

$3F_2 + Ca_3N_2 \longrightarrow N_2 + 3CaF_2$

$2F_2 + Pt \longrightarrow PtF_4$

$3F_2 + 2Au \longrightarrow 2AuF_3$

$3F_2 + U \longrightarrow UF_6$

Gold

$Au + 4HCl + 3HNO_3 \longrightarrow H[AuCl_4] + 3NO_2 + 3HOH$

Hydrogen

$H_2 + Ca \longrightarrow CaH_2$

$6HCl + 6FeCl_2 + KClO_3 \longrightarrow 6FeCl_3 + KCl + 3HOH$

$8HCl + 5FeCl_2 + KMnO_4 \longrightarrow MnCl_2 + KCl + 5FeCl_3 + 4HOH$

$14HCl + 6FeCl_2 + K_2Cr_2O_7 \longrightarrow 6FeCl_3 + 2KCl + 2CrCl_3 + 7HOH$

$4HCl + MnO_2 \longrightarrow MnCl_2 + Cl_2 + 2HOH$

$HCl + NH_3 \longrightarrow NH_4Cl$

$HCl + PH_3 \longrightarrow PH_4Cl$

$6HCl + KClO_3 \longrightarrow 3Cl_2 + KCl + 3HOH$

$6HCl + 5KI + KIO_3 \longrightarrow 6KCl + 3I_2 + 3HOH$

$16HCl + 2KMnO_4 \longrightarrow 2MnCl_2 + 2KCl + 5Cl_2 + 8HOH$

$14HCl + K_2Cr_2O_7 \longrightarrow 2CrCl_3 + 2KCl + 3Cl_2 + 7HOH$

$8HCl + 4SnCl_2 + 2HNO_3 \longrightarrow N_2O + 4SnCl_4 + 5HOH$

$16HCl + 3SnCl_2 + 2K_2CrO_4 \longrightarrow 4KCl + 2CrCl_3 + 3SnCl_4 + 8HOH$

$20HClO_3 + 3As_2S_5 + 24HOH \longrightarrow 20HCl + 6H_3AsO_4 + 15H_2SO_4$

$16HNO_3 + 3Cu_2S \longrightarrow 6Cu(NO_3)_2 + 3S + 4NO + 8HOH$

$6HNO_3 + 5H_2O_2 + 2KMnO_4 \longrightarrow 2Mn(NO_3)_2 + 2KNO_3 + 5O_2 + 8HOH$

$8HNO_3 + H_2S \longrightarrow H_2SO_4 + 8NO_2 + 4HOH$

$HNO_3 + 3FeCl_2 + 3HCl \longrightarrow 3FeCl_3 + NO + 2HOH$

$8HNO_3 + 3Hg \longrightarrow 3Hg(NO_3)_2 + 2NO + 4HOH$

$8HNO_3 + 6KI \longrightarrow 6KNO_3 + 3I_2 + 2NO + 4HOH$

$HOH + NH_4Cl \longrightarrow HCl + NH_4OH$

$HOH + NaCH_3COO \longrightarrow NaOH + CH_3COOH$

$HOH + NaH \longrightarrow NaOH + H_2$

$HOH + NO + NO_2 \longrightarrow 2HNO_2$

$HOH + N_2O_3 \longrightarrow 2HNO_2$

$HOH + K_2O \longrightarrow 2KOH$

$2HOH + Na_2O_2 \longrightarrow 2NaOH + O_2$

$2H_2O_2 \longrightarrow 2HOH + O_2$

$5H_2O_2 + 2KMnO_4 + 3H_2SO_4 \longrightarrow K_2SO_4 + 2MnSO_4 + 5O_2 + 8HOH$

$5H_2S + 2KMnO_4 + 3H_2SO_4 \longrightarrow 2MnSO_4 + 5S + K_2SO_4 + 8HOH$

$2H_2SO_4 + MnO_2 + 2NaBr \longrightarrow Na_2SO_4 + MnSO_4 + Br_2 + 2HOH$

Iodine

$3I_2 + 6NaOH \longrightarrow NaIO_3 + 5NaI + 3HOH$

Iron

$Fe + 4HNO_3 \longrightarrow Fe(NO_3)_3 + NO + 2HOH$

$Fe + 2FeCl_3 \longrightarrow 3FeCl_2$

$Fe(NO_3)_3 + KSCN \longrightarrow [FeSCN](NO_3)_2 + KNO_3$

Lead

$3Pb + 8HNO_3 \longrightarrow 3Pb(NO_3)_2 + 2NO + 4HOH$

$Pb + PbO_2 + 2H_2SO_4 \longrightarrow 2PbSO_4 + 2HOH$

$Pb + NaNO_3 \longrightarrow PbO + NaNO_2$

$3Pb + 4NaNO_3 \longrightarrow Pb_3O_4 + 4NaNO_2$

Lithium

$Li_3N + 3HOH \longrightarrow NH_3 + 3LiOH$

Magnesium

$Mg + 2HOH \longrightarrow Mg(OH)_2 + H_2$

$3MgO + 2Al \longrightarrow 3Mg + Al_2O_3$

$2Mg + CO_2 \longrightarrow 2MgO + C$

Manganese

$MnCl_4 \longrightarrow MnCl_2 + Cl_2$

$MnO_2 + 4HI \longrightarrow MnI_2 + I_2 + 2HOH$

$MnO_2 + H_2C_2O_4 + H_2SO_4 \longrightarrow MnSO_4 + 2CO_2 + 2HOH$

$MnO_2 + H_2O_2 \longrightarrow MnO + HOH + O_2$

$MnO_2 + 2H_2SO_4 + MgBr_2 \longrightarrow MgSO_4 + MnSO_4 + Br_2 + 2HOH$

$4MnSO_4 + 8KOH + O_2 \longrightarrow 2Mn_2O_3 + 4K_2SO_4 + 4HOH$

Mercury

$2HgO \longrightarrow 2Hg + O_2$

$3Hg_2Cl_2 + 2HNO_3 + 6HCl \longrightarrow 6HgCl_2 + 2NO + 4HOH$

Nickel

$Ni + 4CO \longrightarrow [Ni(CO)_4]$

$Ni(OH)_2 + 6NH_3 \longrightarrow [Ni(NH_3)_6](OH)_2$

Nitrogen

$N_2 + 3Mg \longrightarrow Mg_3N_2$

$NH_4Cl + KNH_2 \longrightarrow KCl + 2NH_3$

$2NH_3 + AgCl \longrightarrow [Ag(NH_3)_2]Cl$

$4NH_3 + [Cu(H_2O)_4]SO_4 \longrightarrow [Cu(NH_3)_4]SO_4 + 4HOH$

$NH_4Cl \longrightarrow NH_3 + HCl$

$NH_4Cl + KOH \longrightarrow KCl + NH_3 + HOH$

$10NO + 9H_2SO_4 + 6KMnO_4 \longrightarrow 10HNO_3 + 3K_2SO_4 + 6MnSO_4 + 4HOH$

$(NH_4)_2CO_3 \longrightarrow 2NH_3 + HOH + CO_2$

Oxygen

$3O_2 \longrightarrow 2O_3$

$25O_2 + 2C_8H_{18} \longrightarrow 16CO_2 + 18HOH$

$31O_2 + 4C_6H_7N \longrightarrow 24CO_2 + 14HOH + 2N_2$

$9O_2 + 2C_2H_6S \longrightarrow 4CO_2 + 6HOH + 2SO_2$

$O_2 + 4Fe(OH)_2 + 2HOH \longrightarrow 4Fe(OH)_3$

$2O_2 + SiH_4 \longrightarrow SiO_2 + 2HOH$

Phosphorus

$P_4 + 5O_2 \longrightarrow P_4O_{10}$

$PCl_3 + Cl_2 \longrightarrow PCl_5$

$PH_3 + HCl \longrightarrow PH_4Cl$

Platinum

$PtCl_4 + 2KCl \longrightarrow K_2[PtCl_6]$

Potassium

$6KBr + 6KI + 2KClO_3 + 7H_2SO_4 \longrightarrow 7K_2SO_4 + 2HCl + 3I_2 + 3Br_2 + 6HOH$

$KF + HF \longrightarrow KHF_2$

$KHC_4H_4O_6 + KOH \longrightarrow K_2C_4H_4O_6 + HOH$

$2KMnO_4 + 3HCOOH \longrightarrow 2KHCO_3 + CO_2 + 2MnO_2 + 2HOH$

$64KOH + 27Cl_2 + 2CrI_3 \longrightarrow 2K_2CrO_4 + 6KIO_4 + 54KCl + 32HOH$

$12KOH + 6CoCl_2 + KClO_3 \longrightarrow 3Co_2O_3 + 13KCl + 6HOH$

$K_2CO_3 + K_4[Fe(CN)_6] \longrightarrow 6KCN + FeCO_3$

$3K_2CrO_4 + 2BiCl_3 \longrightarrow 6KCl + Bi_2(CrO_4)_3$

$K_2CrO_4 + Cd(NO_3)_2 \longrightarrow 2KNO_3 + CdCrO_4$

$2K_2CrO_4 + 2HNO_3 \longrightarrow K_2Cr_2O_7 + 2KNO_3 + HOH$

$K_2Cr_2O_7 + 2KOH \longrightarrow 2K_2CrO_4 + HOH$

$2K_2CrO_4 + 16HCl + 3SnCl_2 \longrightarrow 4KCl + 2CrCl_3 + 3SnCl_4 + 8HOH$

$K_2Cr_2O_7 + 14HCl + 3SnCl_2 \longrightarrow 3SnCl_4 + 2KCl + 2CrCl_3 + 7HOH$

$K_3[RhCl_6] + 3K_2C_2O_4 \longrightarrow K_3[Rh(C_2O_4)_3] + 6KCl$

Rhodium

$RhCl_3 + 6NH_3 \longrightarrow [Rh(NH_3)_6]Cl_3$

$RhCl_3 + 3KCl \longrightarrow K_3[RhCl_6]$

Scandium

$2Sc + 3Br_2 \longrightarrow 2ScBr_3$

Silver

$AgCl + 2Na_2S_2O_3 \longrightarrow Na_3[Ag(S_2O_3)_2] + NaCl$

Sodium

$NaHS + NaOH \longrightarrow Na_2S + HOH$

$12NaOH + Fe_4[Fe(CN)_6]_3 \longrightarrow 4Fe(OH)_3 + 3Na_4[Fe(CN)_6]$

$2NaOH + Zn(OH)_2 \longrightarrow Na_2[Zn(OH)_4]$

$2Na_2O + 2H_2SO_4 \longrightarrow 2Na_2SO_4 + 2HOH$

$7Na_2O_2 + 2CrCl_3 + 2HOH \longrightarrow 2Na_2CrO_4 + 6NaCl + 4NaOH + 2O_2$

Sulfur

$5SO_2 + 4HOH + 2KIO_3 \longrightarrow K_2SO_4 + 4H_2SO_4 + I_2$

$SO_2 + NO_2 \longrightarrow SO_3 + NO$

Tin

$SnCl_2 + H_2[PtCl_6] \longrightarrow H_2[PtCl_4] + SnCl_4$

$SnCl_2 + 2K_3[Fe(CN)_6] + 2KCl \longrightarrow SnCl_4 + 2K_4[Fe(CN)_6]$

Xenon

$Xe + F_2 \longrightarrow XeF_2$

$Xe + 2F_2 \longrightarrow XeF_4$

$Xe + 3F_2 \longrightarrow XeF_6$

Zinc

$Zn + 2MnO_2 + 2NH_4Cl \longrightarrow ZnCl_2 + 2MnO(OH) + 2NH_3$

$Zn(OH)_2 + 4NH_3 \longrightarrow [Zn(NH_3)_4](OH)_2$

$Zn(OH)_2 + 2KOH \longrightarrow K_2[Zn(OH)_4]$